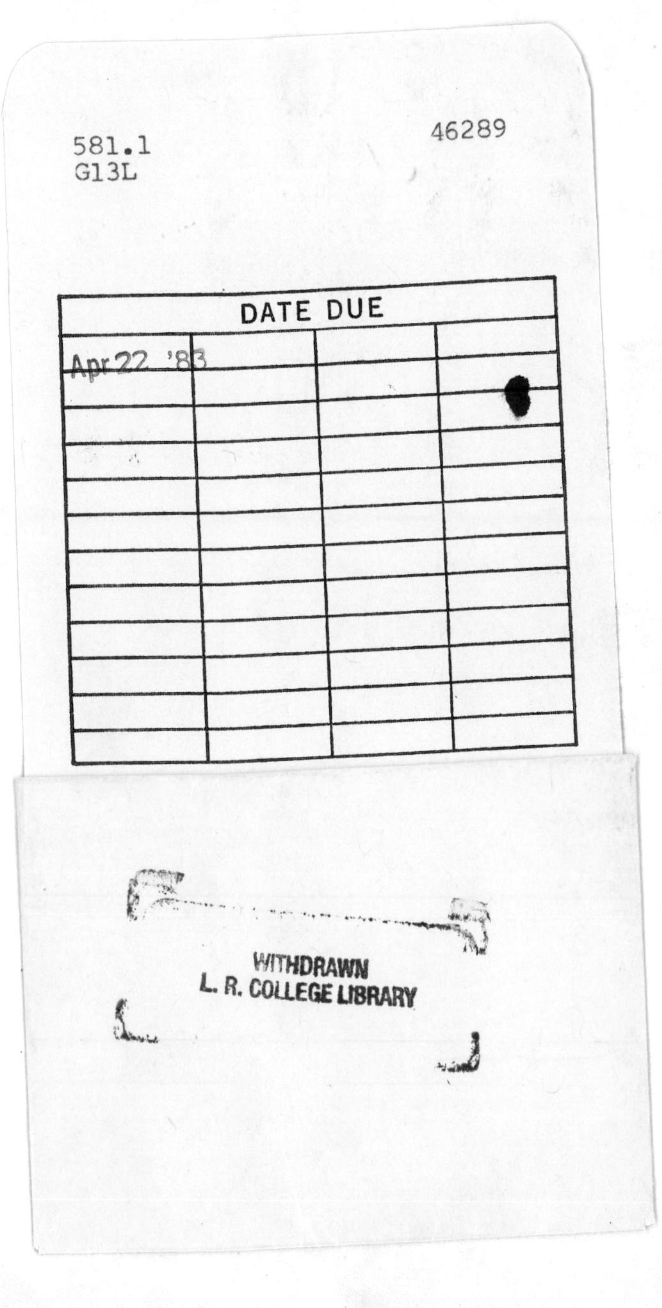
581.1
G13L
46289
DATE DUE
Apr 22 '83
WITHDRAWN
L. R. COLLEGE LIBRARY

PRENTICE-HALL FOUNDATIONS OF MODERN BIOLOGY SERIES

The Cell, *by Carl P. Swanson*

Cellular Physiology and Biochemistry, *by William D. McElroy*

Heredity, *by David M. Bonner*

Adaptation, *by Bruce Wallace and A.M. Srb*

Animal Growth and Development, *by Maurice Sussman*

Animal Physiology, *by Knut Schmidt-Nielsen*

Animal Diversity, *by Earl D. Hanson*

Animal Behavior, *by Vincent Dethier and Eliot Stellar*

The Life of the Green Plant, *by Arthur W. Galston*

The Plant Kingdom, *by Harold C. Bold*

Man In Nature, *by Marston Bates*

The Life of the Green Plant

FOUNDATIONS OF MODERN BIOLOGY SERIES

The Life of the Green Plant

ARTHUR W. GALSTON

Yale University

Prentice-Hall, Inc.

ENGLEWOOD CLIFFS, NEW JERSEY

The Life of the Green Plant

Arthur W. Galston

*Printed in the
United States of America
Library of Congress
Catalog Card Number 61–8697*

Third printing..................November, 1961

PRENTICE-HALL FOUNDATIONS OF MODERN BIOLOGY SERIES

William D. McElroy and Carl P. Swanson, *Editors*

Design by Walter Behnke

Drawings by Felix Cooper

C

To **Dale, Bill, and Beth**

Foundations of Modern Biology Series

The science of biology today is *not* the same science of fifty, twenty-five, or even ten years ago. Today's accelerated pace of research, aided by new instruments, techniques, and points of view, imparts to biology a rapidly changing character as discoveries pile one on top of the other. All of us are aware, however, that each new and important discovery is not just a mere addition to our knowledge; it also throws our established beliefs into question, and forces us constantly to reappraise and often to reshape the foundations upon which biology rests. An adequate presentation of the dynamic state of modern biology is, therefore, a formidable task and a challenge worthy of our best teachers.

The authors of this series believe that a new approach to the organization of the subject matter of biology is urgently needed to meet this challenge, an approach that introduces the student to biology as a growing, active science, and that also *permits each teacher of biology to determine the level and the structure of his own course.* A single textbook cannot provide such flexibility, and it is the authors' strong conviction that these student needs and teacher prerogatives can best be met by a series of short, inexpensive, well-written, and well-illustrated books so planned as to encompass those areas of study central to an understanding of the content, state, and direction of modern biology. The FOUNDATIONS OF MODERN BIOLOGY SERIES represents the translation of these ideas into print, with each volume being complete in itself yet at the same time serving as an integral part of the series as a whole.

Contents

1 The Green Plant in the Economy of Nature

The story of life on earth, like the story of the earth itself, begins with the sun. Except for man's recent utilization of atomic power, the sun is the sole source of energy for almost all forms of living matter. All functioning machines require some source of energy to make them go: a watch uses the energy of a coiled spring; a hydroelectric plant employs the kinetic energy of falling water; an automobile runs on gasoline, by releasing its chemical energy through the process of oxidation. Similarly, all living cells obtain their energy from oxidizable fuels called foods. Food molecules are chemically very diverse, but we can get an idea of their nature by examining the most important one, the simple sugar, glucose, $C_6H_{12}O_6$, which contains 6 carbon atoms, 12 hydrogen atoms, and 6 oxygen atoms. This substance, the basic food molecule found in almost every living cell, is derived directly from the energy of the sun through the operation of the photosynthetic machinery of a living green cell (Fig. 1).

When a green plant grows, therefore, it is in fact tapping solar energy. Since man consumes either green plants or creatures that eat green plants, he, too, is indirectly drawing on solar energy. Even the gasoline-powered automobile utilizes "fossilized" solar energy captured in photosynthesis by organisms that died millions of years ago. Were there no green plants to function as solar energy converters, practically all life on earth would cease. The only exceptions

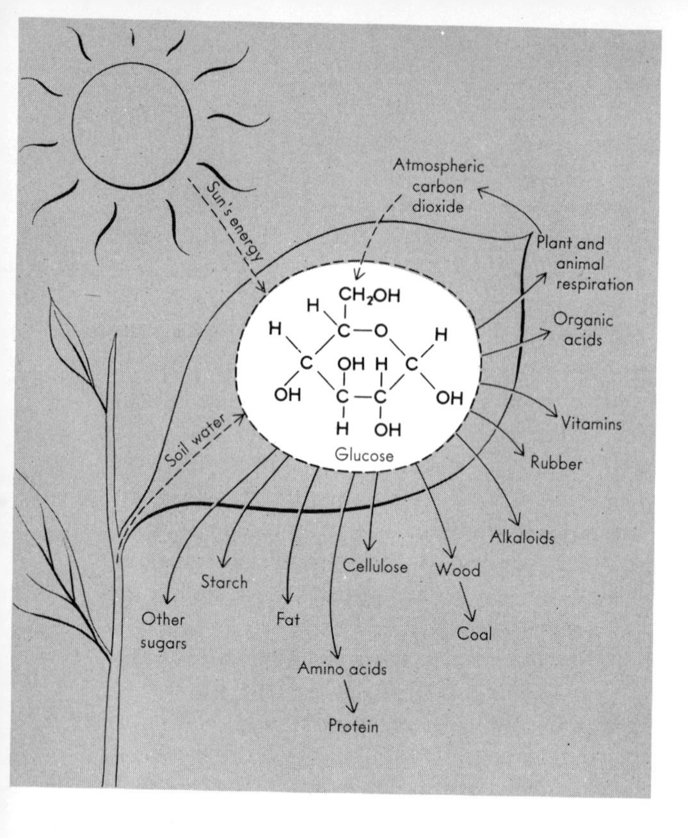

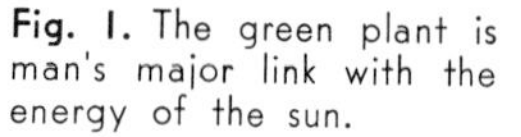

Fig. 1. The green plant is man's major link with the energy of the sun.

might be certain bacteria which derive their energy by the oxidation of unusual substrates such as ferrous iron. Even these organisms, however, are probably indirectly dependent on solar energy, and, in any event, they do not bulk large in the over-all inventory of life on earth.

The Sun as a Thermonuclear Device

Since the end of World War II, man has become increasingly familiar with the tremendous quantities of energy that can be released through the interactions of subatomic particles. Although the atomic age originated in fission-type reactions, in which large atoms like uranium are degraded to smaller atoms and subatomic particles, much of the emphasis in nuclear energy research today centers around the fusion type of reaction, in which small units, such as protons, are fused into larger units, such as alpha particles. This reaction is the basis for the hydrogen bomb and is also being studied in the attempt to achieve controlled thermonuclear fusions as a source of energy for industrial purposes.

The sun, in fact, is a kind of hydrogen bomb. It is a thermonuclear device in which four hydrogen atoms of approximately mass 1 are fused into helium of approximately mass 4 through a complicated series of reactions. The over-all equation may be symbolized in this way:

$$4H \longrightarrow He$$

Actually, each of the four hydrogen atoms participating in the thermonuclear fusion has a mass of 1.008, while the helium atom resulting from their fusion has a mass of 4.003. Since more mass is going into the reaction ($4 \times 1.008 = 4.032$) than is coming out ($1 \times 4.003 = 4.003$), the equation is unbalanced. The difference in mass (0.029 units of mass) is converted into energy, according to the Einstein equation ($E = Mc^2$), in which E is the energy produced, in ergs, M is the mass of matter transformed, in grams, and c is the velocity of light (3×10^{10} cm/sec). Although the erg is a very small unit (it takes more than 40 million to make one calorie), the equation shows that large quantities of energy are released by the conversion of very small amounts of mass into energy. It has been estimated that deep within the sun some 120 million tons of matter vanish every minute, being converted into the tremendous quantities of energy radiated into space.

Of this solar radiation, the earth's surface receives annually approximately 5.5×10^{23} calories, or about 100,000 cal/cm^2/yr. Of this total, about one-third is expended in evaporating water, leaving on the order of 67,000 cal/cm^2/yr. for photosynthesis and other purposes. Every year photosynthesis converts 200 billion tons of carbon from atmospheric carbon dioxide to sugar; this amount is about 100 times the combined mass of all the goods that man produces in a year. Despite the fact that photosynthesis is the most extensive chemical process on earth, the green plant is relatively inefficient in utilizing this radiant energy. Annual photosynthesis over the earth averages only about 33 cal/cm^2, which means that photosynthesis converts only about 1/2000 of the energy that is available to it. This figure does not give us an accurate picture of the efficiency, however, for there are no plants at all where much of the radiant energy falls on the earth. If proper correction is made for the radiation absorbed by the green-plant surface of the earth, the efficiency of the world's photosynthesis rises to perhaps several per cent.

Radiant Energy

As hydrogen is converted to helium in the solar thermonuclear furnace, many kinds of radiations are produced. Although different in many respects, these radiations are part of a continuous spectrum of energy, in which the various kinds of radiation are characterized by different wavelengths (Fig. 2). To express these wavelengths numerically, it is convenient to use *millimicrons* as the fundamental unit (one micron is one millionth of a meter; a millimicron is a billionth of a meter). The visible spectrum extends from approximately 400 to about 700 millimicrons. Some people can see farther than this into the ultraviolet

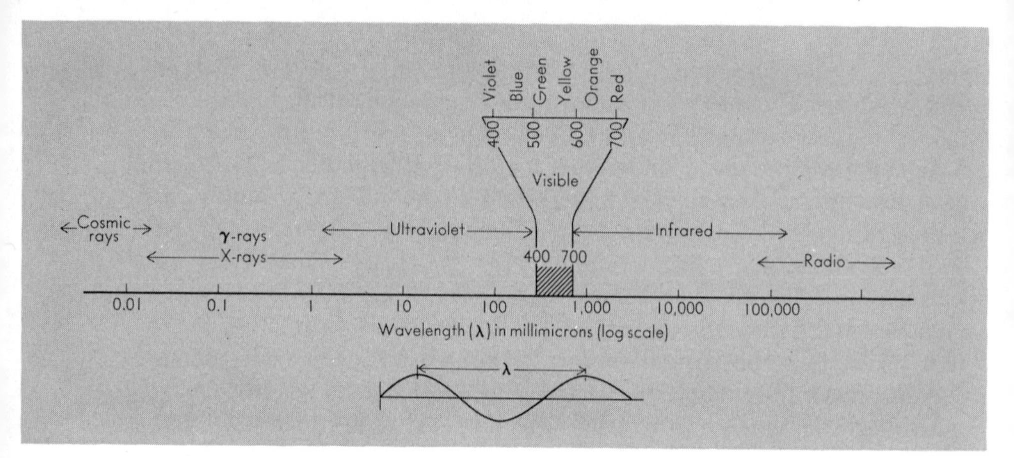

Fig. 2. The spectrum of radiant energy plotted on a logarithmic wavelength scale.

and the infrared, but these limits can be taken as a reasonable average. Four hundred millimicrons represent the blue-violet end of the spectrum and 700 the red end—the colors of the spectrum ranging in order from violet, blue, green, yellow, and orange, to red. Curiously, a plant is sensitive to almost exactly the same range of radiation as is the human eye, although certain kinds of bacteria can utilize infrared radiations we cannot see.

Early in the twentieth century the German physicist, Max Planck, established that the energy of radiation is contained in packets called *quanta* and that the energy content of these quanta is directly proportional to the frequency of the radiation. In other words:

$$\underset{\text{Energy of the quantum}}{E} = \underset{\text{Planck's constant}}{h} \cdot \underset{\text{Frequency of the radiation}}{\nu}$$

Since all the radiations travel at the same speed (3×10^{10} cm/sec) and since frequency times wavelength equals velocity of light, the frequency can be deduced from a known wavelength and vice versa. Thus the longer the wavelength of light the lower the frequency and the less energetic the quantum.

The complex of radiations from the sun is greatly altered by the time it reaches the earth. For example, the ozone (O_3) of the atmosphere absorbs strongly in the ultraviolet region. This is fortunate, since the undiminished ultraviolet radiation would cause severe damage to living terrestrial systems. The infrared (heat) radiations are largely absorbed by water vapor and, to some extent, by the small amount of carbon dioxide in the atmosphere. This, too, helps keep the earth's temperature in the proper range for living systems. The energy that finally hits the earth is mostly in the visible and infrared range, but also extends somewhat into the ultraviolet. These radiations that penetrate the atmosphere constitute the basis for the energetics of all living systems on earth. It is the green plant that is able to store a portion of this radiant energy in the process of photosynthesis.

Human Populations and Food Supply

We live in unusual times. The population of the earth, now around 2.9 billion people, is increasing at an unparalleled rate of about 1.7 per cent per annum (Fig. 3). At this rate, forty-seven million additional consumers of food are added to the earth every year—about the equivalent of the population of France or Italy. Thus the *daily* net increase (births minus deaths) is almost 100,000 people, or *more than one extra mouth to feed every passing second!* Furthermore, as the total number of people on earth increases, the annual population increment goes up. As public health measures are extended and improved, consequently lowering the death rate, the *rate* of increase in population will *itself* rise. We can expect the human population, therefore, to just about double every 40 years. Should this continue for another millenium, the weight of the people on the earth would approximately equal the weight of the planet itself. Obviously something must change before the "population bomb" engulfs us all.

Since all animals, including man, are dependent for their fuel on the solar energy trapped by green plants, any calculation of how many people we can comfortably support on the earth ultimately depends on the amount of energy that can be trapped by photosynthesis. To what

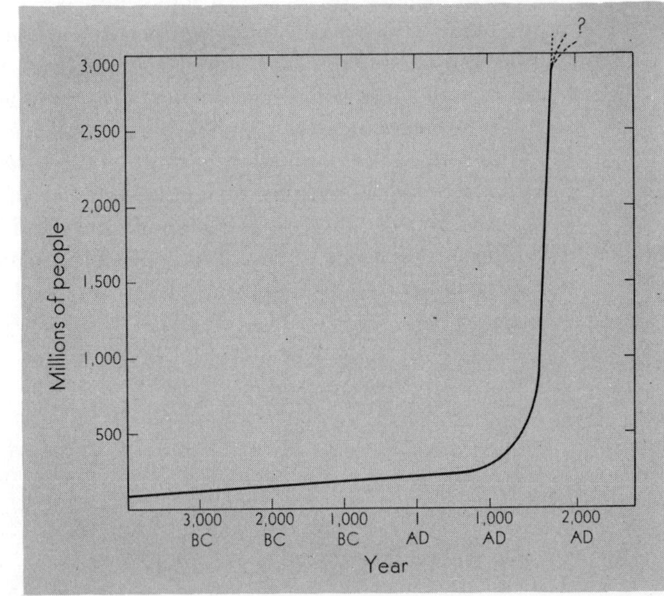

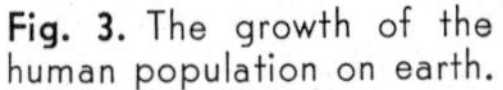
Fig. 3. The growth of the human population on earth.

extent can the two hundred billion tons of carbon stored per year in photosynthesis be increased? Even Herculean efforts to expand the acreage of agriculturally productive land can do no more than double the area under cultivation, but since the best land is already in use, this increased effort will not double productivity. Most estimates of photosynthetic productivity reveal that the waters of the earth contribute markedly to the total; at least half, and possibly up to 80 per cent of the photosynthesis on earth occurs in marine and fresh-water environments. Can we, then, "farm" the seas, or perhaps cultivate algae of various kinds for food in great "tank cultures" of highly fertilized aqueous media? Although at present economically unfeasible, such a system may one day be necessary, when the world is bulging with hungry people. In this new kind of agriculture, botanical "know-how" will obviously play a large role.

Another way to increase food production is to improve the plant itself. Much has been accomplished in this field already by the development of scientific agriculture, to which botanists have contributed greatly. The geneticist has given us increasingly better kinds of plants to work with; the plant physiologist has taught us to care for the nutritional needs of the plant and to alter its growth habit by specific chemical treatments; the plant pathologist has shown us how to ward off insect and fungus pests; and the soil scientist has demonstrated how to enrich and preserve the complex soil environment of pulverized rock, organic matter, and organisms. Some day, perhaps, we may understand the mechanism of photosynthesis so well that we will be able to control and improve its efficiency within the plant, or even duplicate it efficiently outside the living cell.

Even if the food productivity of the earth is increased, it will be more than negated by an unrestricted increase in the human population. Why struggle to double food productivity when in 40 years the gain will be completely erased by a doubling of the number of mouths to feed? Clearly, man must some day decide what number of people can comfortably be accommodated on the earth's surface. Then means must be found to restrict the population to that number. Although such a proposal raises a tremendous number of religious, political, and sociological questions, human populations will eventually have to be limited in some way. The alternatives seem to be either a peaceful, planned, voluntary limitation or a violent and chaotic limitation, imposed by starvation, disease, or war.

2
The Green Plant Cell

Scientists generally try to reduce the problems in their field to the simplest possible terms. Biologists have, for more than a century, agreed that the cell is the basic structural and functional unit in all organisms. For unicellular forms, of course, the cell *is* the organism; multicellular forms introduce further problems of organization and differentiation, and of cooperation and competition among cells. Although no responsible biologist will claim that an understanding of the cell is the equivalent to complete knowledge of the organism, the cell is certainly a logical starting place for a meaningful study of an organism. Let us therefore review the general characteristics of all cells, and then investigate in detail the unique characteristics of the cells of higher green plants.

Cells vary tremendously in size, from bacterial cells less than a micron in diameter to elongated cells several millimeters in length. Even a relatively small bacterial cell contains on the order of 10^{12} molecules. The tremendous complexity of this basic biological unit should make it clear why the precision and methods in biological research are quite different from those in physics or chemistry, where the units may be as elementary as a single proton or an individual quantum.

A plant cell grown in isolation generally assumes a spherical shape (Fig. 4), but when surrounded by other cells it is polyhedral. A cell from the elongating zone of a stem or root has the approxi-

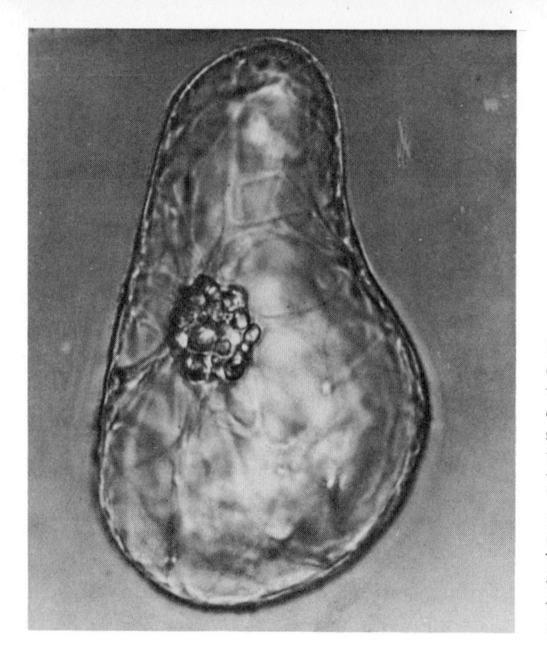

Fig. 4. A large single isolated cell of pea grown in hanging drop culture. Note the cell wall, the large central nucleus surrounded by starch grains, and the cytoplasmic threads suspending it in a large vacuole. (From John G. Torrey, "Experimental Modification of Development in the Root," in *Cell, Organism and Milieu,* edited for The Society for the Study of Development and Growth by Dorothea Rudnick. Copyright 1959 The Ronald Press Company.)

mate shape of a box, and is about 50 μ long by 20 μ wide by 10 μ deep, with a volume of about 10,000 μ^3. One hundred million such cells, if tightly packed, would fit in a volume of 1 cubic centimeter. The complex and highly differentiated structure of a plant cell is composed of three main regions: (1) the *cell wall,* a relatively rigid, presumably non-living, chemically complicated substance that is secreted by the rest of the cell, (2) the *protoplast,* or living portion of the cell, which is completely surrounded by differentially permeable membranes, and (3) *vacuoles,* the nonliving storage tanks that contain an aqueous solution of inorganic salts and various organic molecules produced by the metabolic activities of the cell. The wall of the cell serves as the plant's skeleton, by contributing to the rigidity and form of the plant body. The vacuoles function as a sort of excretory system, since material deposited within them is effectively removed from the scene of active chemical transformations in the cell. It is to the protoplast, then, that we must look for the seat of the ceaseless activities that characterize the highly organized and dynamic state we call life.

The Nucleus and Ribosomes

Within the last several decades, biologists have become reasonably certain of the structure, chemistry, and functional significance of the major constituents of the cell's protoplast. The major organelle of the cell, in size and importance, is the *nucleus* (Fig. 5). This globose body, generally about 5–10 microns in diameter, contains the basic genetic information of the cell, in the form of long strands of a complex chemical material called *deoxyribose nucleic acid,* or DNA. The DNA is usually

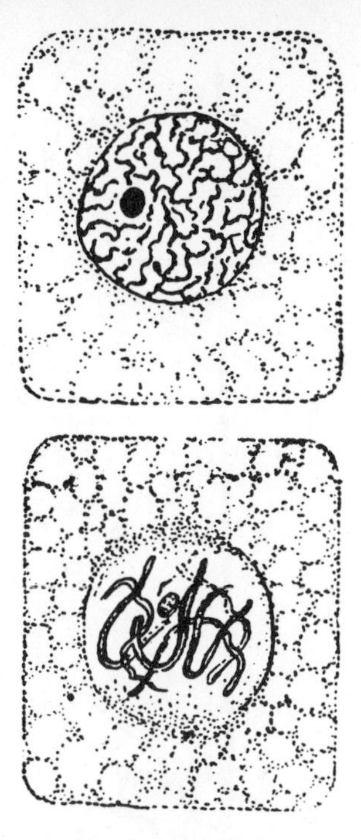

Fig. 5. The nucleus in a nondividing (top) and a dividing cell (bottom).

seen in the non-dividing cell in what looks like a network of *chromatin*, so called because of its affinity for certain highly colored dyes. When the cell is about to divide into two cells, it becomes clear that the chromatin is really present in rod-like bodies called *chromosomes*, of which each cell has a basic and fixed number. For example, almost every cell in the pea plant has 14 chromosomes, divided into two groups of 7 chromosomes, one group being derived from each of the two parents. Similarly, in man, the basic number is 46, of which half comes from the mother and half from the father.

The full double complement of chromosomes is referred to as the *diploid* ($2n$) number, and the basic number obtained from each parent the *haploid* (n) number. All the cells of the higher plant body are at least diploid, although the sex cells, found in the mature pollen grains and embryo sac in the ovules of the flower, contain the haploid number. In the reproduction process, the haploid number is produced by *reduction division* or *meiosis*, carried out in *spore mother cells*. When the haploid sex cells fuse to produce the *zygote*, the diploid number is restored. Thus, in terms of chromosome content and DNA, the higher plant goes through a cycle in which diploidy alternates with haploidy, and in which the fusion of haploid cells of different genetic origins to form new diploid individuals provides an opportunity for the production of novel types of organisms.

Until recently, all the cells of the plant body were thought to be diploid, but new evidence has revealed the existence of pockets of cells with multiples of the diploid condition, such as, $4n$, $6n$, $8n$, etc. These polyploids probably arise as a result of nuclear divisions which are not followed by separation into two cells. The ploidy of cells can now be controlled by the use of natural and synthetic chemicals of various types.

The molecules of DNA, found in the chromosomes, are believed to be in the form of double strands of intertwined helices (Fig. 6). The individual molecules, which are very long, are linear aggregates of four basic building blocks, called *nucleotides*. The order in which the nucleotides occur in the chain probably determines the genetic information carried by the chain. Thus, if the four nucleotides are represented by *A, B, C,* and *D,* a linear aggregate in an *-ABCD-* order will constitute different genetic information from -*ACBD*- or -*ADBC*- or -*ABBACDCD*- chains. Our inheritance, therefore, and that of each organism, probably consists in its essence of some repeated pattern of nucleotide units in the DNA molecules of the chromosomes contained in our nuclei.

For most of the life of the cell, the nucleus is separated from the rest of the protoplast, called *cytoplasm,* by a nuclear membrane. This membrane is revealed by electron microscopy to be double in nature and to possess both pores and projections into the cytoplasm. During the nuclear division *(mitosis)* preceding cell division, the nuclear membrane appears to break down completely, leaving the nuclear material free to mix with the rest of the protoplast. At this point, the *nucleolus,* one or several of which are found in every nucleus, disappears completely as an organized body and presumably dissolves into the cytoplasm. The nucleolus is formed in the nucleus at a particular locus on one of the chromosomes and is mainly composed of a material called *ribose nucleic acid* (RNA). RNA differs chemically from DNA in that each nucleotide unit contains the sugar called *ribose* instead of the analogous sugar found in DNA, which is called *deoxyribose* because it lacks one oxygen. RNA, like DNA, probably consists of long, linear aggregates of repetitive nucleotide patterns, and, in fact, the nuclear DNA is thought to impress on the RNA of the cell some basic pattern that is char-

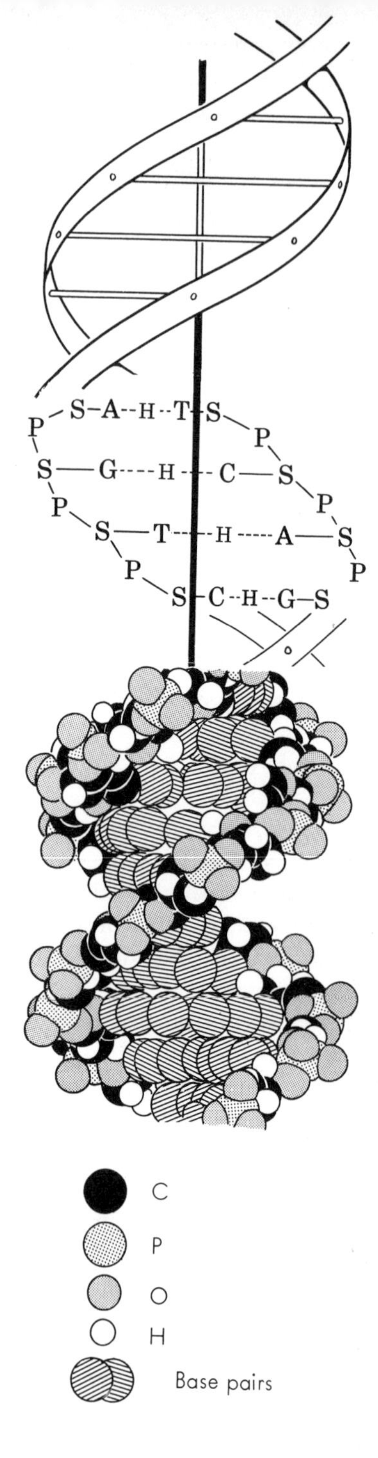

Fig. 6. The spiral structure of the DNA molecule. The twin spirals are long, linear, polynucleotide chains. The letters *P* and *S* denote the phosphate-sugar backbone of each spiral; the letters *A, T, G,* and *C* represent adenine, thymine, guanine, and cytosine, the four basic substances of the nucleotide units; *H* denotes a hydrogen bond between these base pairs, holding the spirals together. (From Carl Swanson, *The Cell*. Englewood Cliffs, N.J.: Prentice-Hall, 1960.)

acteristic of that type of cell. Biologists now believe that all biological units capable of self-replication (chromosomes, viruses, etc.) depend on either DNA or RNA for the transfer of basic information.

We now have very good evidence that RNA is required for the synthesis of the proteins of the cell. Most researchers believe that the specificity of a protein, which results from the particular sequence of amino acids in the protein, is determined by some structural characteristic of the RNA molecules of the cell. Ultimately, this special characteristic is traceable to the sequence of nucleotide elements in the RNA molecules, which in turn is probably determined by the nuclear DNA. Thus the DNA in the genes of the nucleus regulates the specificity of the protoplasmic proteins through the transfer of specificity from DNA to RNA to protein. We do not know the exact mechanism by which RNA controls the nature of the amino acid sequence of the protein molecule, but it is logical to assume that the sequence of nucleotides in RNA produces particular surface characteristics in the RNA molecule, and that certain amino acids fit into these specific niches in the RNA chain.

Large protein molecules are synthesized from the component amino acids in various parts of the cytoplasm, including the *ribosomes,* small spherical bodies only a fraction of a micron in diameter. The ribosomes seem to be localized swellings of a much-branched system of canal-like double membranes called the *endoplasmic reticulum* (Fig. 7). These canals emanate from the double membrane of the nucleus and reach into all corners of the cytoplasm. The ribosomal globules can be centrifuged in bulk out of cells disrupted in a homogenizer and can synthesize proteins from amino acids when supplied with appropriate RNA. Although the ribosomes are probably the main synthesizers of cellular protein, other cellular organelles, such as mitochondria and chloroplasts, apparently also serve this function.

Many of the proteins synthesized by ribosomes and other cellular organelles regulate the rate of chemical reactions occurring in the cell. These proteins possessing specific catalytic properties are called *enzymes.* Their role is a crucial one, for they determine the ultimate nature of the cell by controlling the chemical reactions that synthesize the cell's constituents. In certain species, for example, the genetic difference between red and white flowers is due to a single pair of genes. Chemically, the difference stems from the fact that the petal cells of the red variety possess an enzyme which can convert a colorless precursor substance to a red pigment, while the white form is unable to carry out this transformation. The DNA of the nuclear genetic material, therefore, determines the color of the petals by regulating the synthesis of the cytoplasmic enzyme that produces a colored substance from a colorless one (Fig. 8). This control is presumably transferred to the cytoplasm through RNA made in the nucleus.

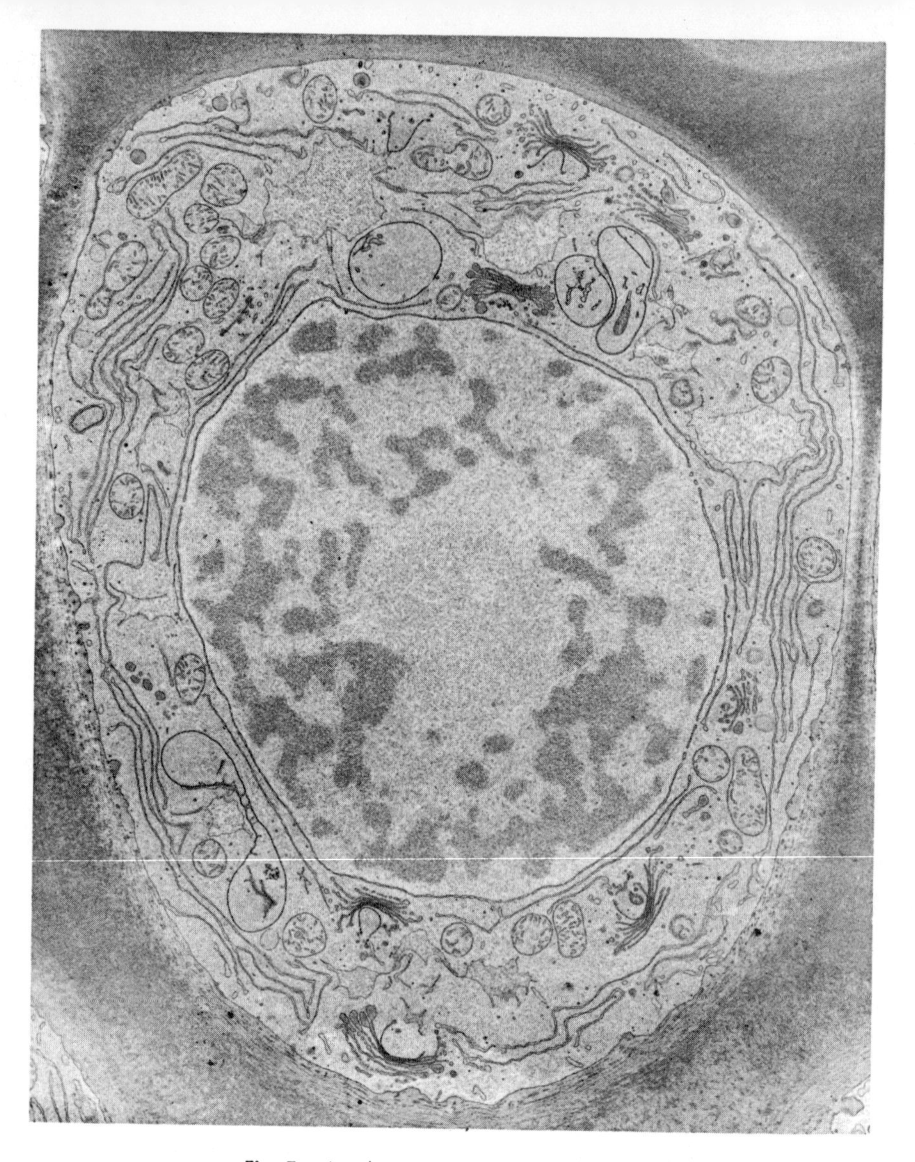

Fig. 7. An electron microscopic view of a cell of a maize root tip. This is a transverse section, taken 75 μ from the apex. The large central body is the *nucleus;* note the darker *chromatin* areas and the numerous pores in the double nuclear membrane. The elongated canals running through the cytoplasm are portions of the *endoplasmic reticulum,* which represents projections of the nuclear double membrane. The numerous club-shaped or rounded bodies with internal projections are the *mitochondria* with their *cristae.* The close groups of short canals with vesicular ends are *Golgi bodies;* the lighter stippled-appearing areas in the cytoplasm are the *vacuoles.* (Courtesy W. G. Whaley and the University of Texas Electron Microscope Laboratory.)

There are thousands of enzymes in any cell, and each controls one chemical reaction or a group of related chemical reactions. Many of these enzymes have been extracted from the cell and then purified,

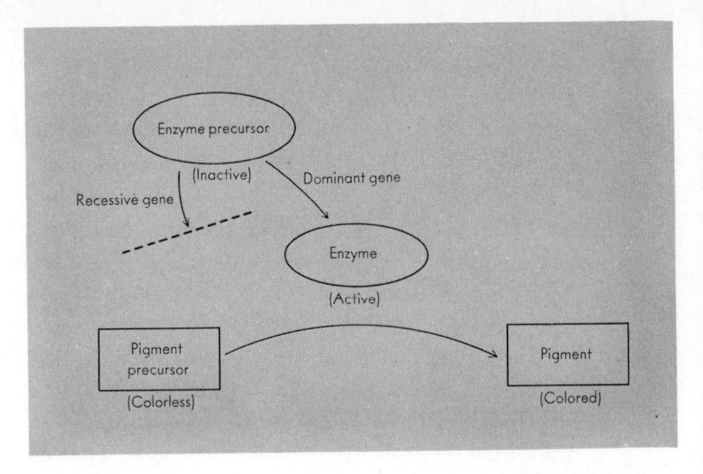

Fig. 8. Genes may control the synthesis of specific enzymes which in turn control cell chemistry.

isolated, and finally crystallized. They are composed largely or exclusively of protein. Certain enzymes are entirely protein, while others consist mainly of protein (the *apoenzyme*) to which is attached a smaller molecule (the *coenzyme*) (Fig. 9). In such instances, neither the apoenzyme nor the coenzyme alone can function catalytically; if the two are dissociated, catalytic activity ceases, and if they are recombined, the activity can be completely restored.

Among the kinds of molecules that function as coenzymes are (1) such *metals* as iron, manganese, copper, zinc, molybdenum, and magnesium, and (2) such *vitamins* as thiamin, riboflavin, nicotinic acid, and pyridoxine. With both metallic and vitamin-type coenzymes, the active coenzymatic unit may be somewhat more complicated than the single metal or vitamin; for example, iron may be present as *heme,* a complex organic molecule with an iron center, and nicotinic acid may be present as a triphosphorylated derivative. Certain enzymes, such as the metalloflavoproteins, have more than one type of coenzyme. In one particular

Fig. 9. Some enzymes consist of a large protein *apoenzyme* and small *coenzyme*. Only the apoenzyme plus coenzyme complex is active.

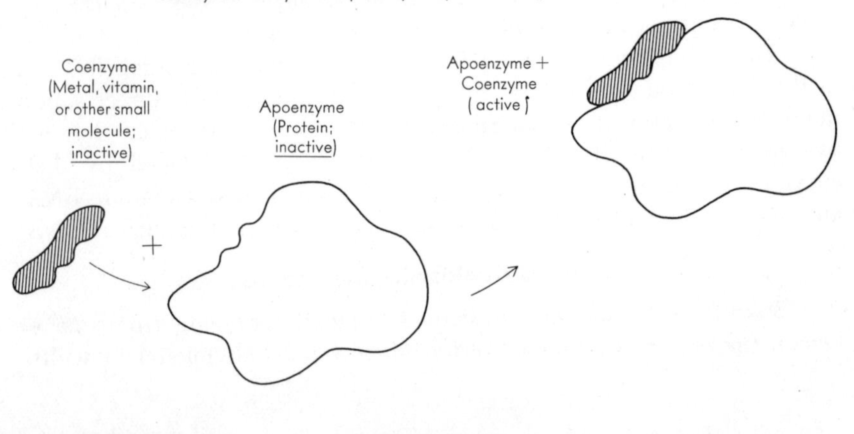

metalloflavoprotein called aldehyde oxidase, the active enzymatic molecule contains, in addition to the basic structural protein, free iron, iron in heme, and riboflavin in a complex form called flavin-adenine dinucleotide. All these coenzymes are required for activity, and they must be attached at the correct position on the protein if they are to be effective.

Enzymes catalyze a wide variety of chemical reactions, including synthetic, degradative, hydrolytic, oxidative, and reductive reactions. They also catalyze the transfer of groups such as amino, methyl, phosphate, etc. In general, one enzyme with a specificity that is ultimately gene-determined catalyzes only one reaction or type of reaction. All enzymes seem to function by first forming a chemical complex with the substances upon which they act. The enzyme-substrate complex then undergoes some internal rearrangements that cause alterations of the substrates and the eventual release of the products of the reaction. For example, suppose that two small molecules, *A* and *B*, slowly unite to form the larger molecule, *AB*, and that this reaction is catalyzed by the enzyme, *E*. The over-all reaction,

$$A + B \xrightarrow{E} AB$$

can be shown to consist of the following steps:

$$\begin{aligned} E + A &\longrightarrow EA \\ EA + B &\longrightarrow EAB \\ EAB &\longrightarrow E + AB \end{aligned}$$

If these equations are added together, the net reaction is A plus $B \longrightarrow AB$. Note that the enzyme does not appear as a constituent of the reaction. The regeneration of the enzyme in the final step accounts for its catalytic role in the over-all reaction, and only small amounts of enzyme are required to produce large total changes in substrate and product level. Very small quantities of certain mineral elements and vitamins are of vital importance in physiological processes because they serve as coenzymes for specific enzymatic molecules.

The enzymes of the cell, located in the various cell organelles and in the non-particulate area of the cytoplasm, are thus the direct superintendents of the cell's chemical machinery. All cells are what they are by virtue of their chemistry; their chemistry is determined by their enzymes; the nature of the enzymes is determined by cytoplasmic RNA; and the specificity of the RNA is in turn determined by the nuclear DNA.

Other Cytoplasmic Organelles

Mitochondria are sausage-shaped organelles ranging from one to several microns in length and approximately one-half micron in width.

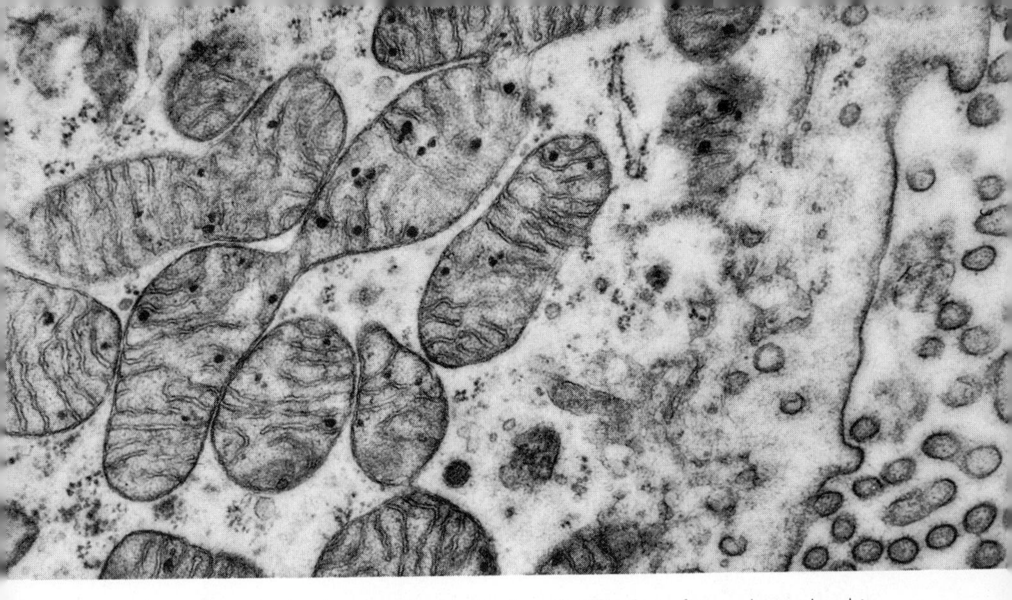

Fig. 10. An electron microscopic view of a thin slice of several mitochondria (left). Note the membrane and *cristae*. (Courtesy G. E. Palade.)

They have a double membrane, and the inner one is convoluted into numerous plate-like *cristae* (Fig. 10). The mitochondria are capable of oxidizing various organic molecules, thus releasing energy in the process. This is believed to be their main function in the cell. Like the ribosomes, they can be isolated in bulk by centrifugation of a cell homogenate, and then fed the appropriate organic substances. During the mitochondrial oxidations, the energy is stored in the form of special energy-rich phosphate bonds of adenosine triphosphate (ATP) (Fig. 11). Since the cell can utilize the energy in the bonds to perform work of various kinds, the continuing mitochondrial oxidations supply the cell with its required "energy currency" of ATP. The typical cell contains hundreds of mitochondria, distributed throughout the cytoplasm. Cells

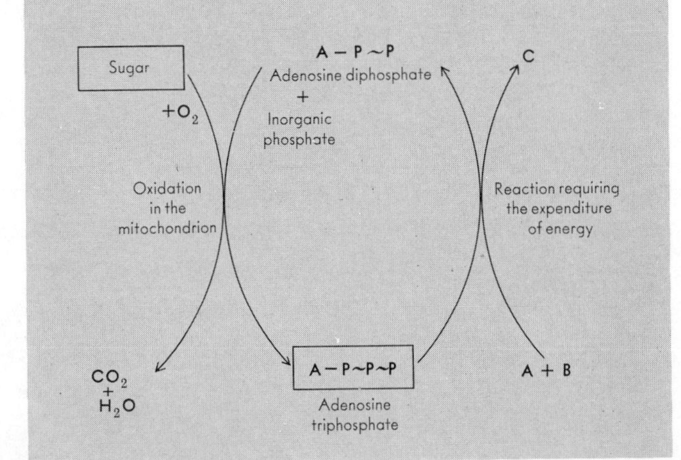

Fig. 11. In the oxidation of sugar by the cell, energy is stored in the form of special "energy-rich" phosphate bonds in the molecule of adenosinetriphosphate (ATP). The ordinary phosphate bond is represented by —P, the energy-rich bond by ∼P. The ATP can then be used to drive reactions requiring the expenditure of energy, such as the union of two small molecules (A, B) to form a larger one (C).

that are especially active metabolically contain mitochondria that are larger and more numerous than the average.

The *Golgi apparatus,* which consists of a group of bladder-like membranes, occurs widely in plant and animal cells. Its function is completely unknown and awaits further investigation.

So far we have discussed the organelles found in all cells. What about those restricted to green plant cells? Plant cells are unique in that they possess a variety of globular cytoplasmic bodies called *plastids.* Of foremost importance in the green plant cell is the *chloroplast* (Fig. 12), the seat of photosynthetic activity of the green cell and site of all the chlorophyll and accessory pigments associated with photosynthetic activity. In higher plant cells, the chloroplast is a disc-shaped body roughly 5–8 μ in diameter by about 1 μ in width. Approximately 50 such bodies are present in each cell, and, as far as we know, each chloroplast is derived from a proplastid transmitted maternally. Proplastids are presumably able to replicate by some sort of division process, thus increasing their number in the cell, but mature chloroplasts are apparently incapable of such replication.

Electron microscopy of chloroplasts has revealed them to be constructed of parallel double layers running the length of the plastid. In places, these layers are thicker than their usual cross-sectional dimension, and when several thicken at the same point, they form *grana,* the areas of the chloroplast in which the chlorophyll is found. The non-green areas of the plastid are called *stroma,* and the entire plastid is surrounded by a membrane with differentially permeable properties.

Fig. 12. (Left) An electron microscopic view of the structure of a thin slice of corn chloroplast. Note the *lamellae* and *grana.* (Right) A higher-powered view. (Courtesy S. Klein.)

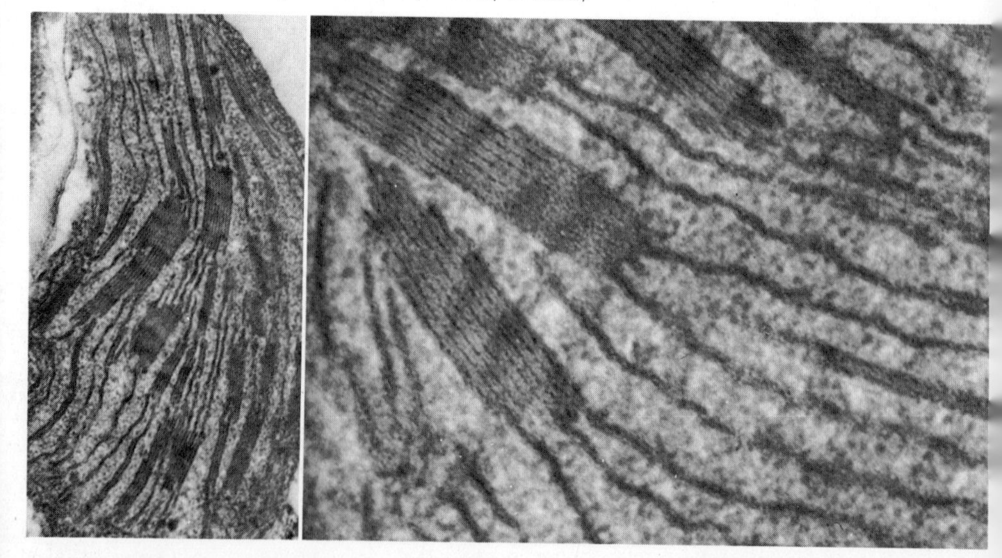

In the flowering plants, or angiosperms, the mature chloroplast will not develop from the proplastid unless the plant is illuminated, while in certain gymnosperms the transformation can be accomplished in total darkness. The two groups apparently differ in their ability to transform the pigment protochlorophyll to chlorophyll, a change that involves the addition of two hydrogen atoms to the protochlorophyll molecule (Fig. 13). In the angiosperms, this step is dependent on light, but in the gymnosperms it can proceed in the dark.

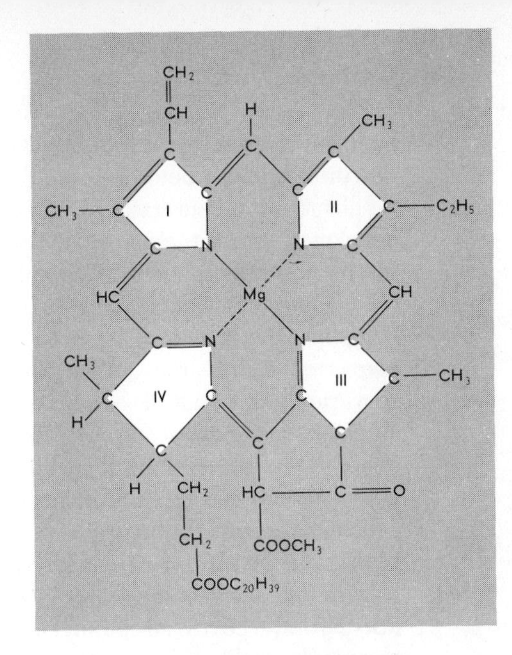

Fig. 13. The chemical structure of the chlorophyll *a* molecule. Note the magnesium center and the four pyrrole rings (I-IV) attached to it. This structure resembles *heme*, which has iron at the center bound to four pyrrole rings.

In etiolated (dark-grown) angiosperm leaves, the proplastids fail to develop to maturity. They are smaller than chloroplasts and do not have the layered structure characteristic of the chloroplast. The proplastids possess instead a sort of paracrystalline center of canals that, under proper stimulation by light, develop into layers (Fig. 14). In potentially green angiosperm cells which

Fig. 14. An electron microscopic view of an etiolated proplastid. Note the central prolamellar body and rudimentary unoriented lamellae. (Courtesy S. Klein.)

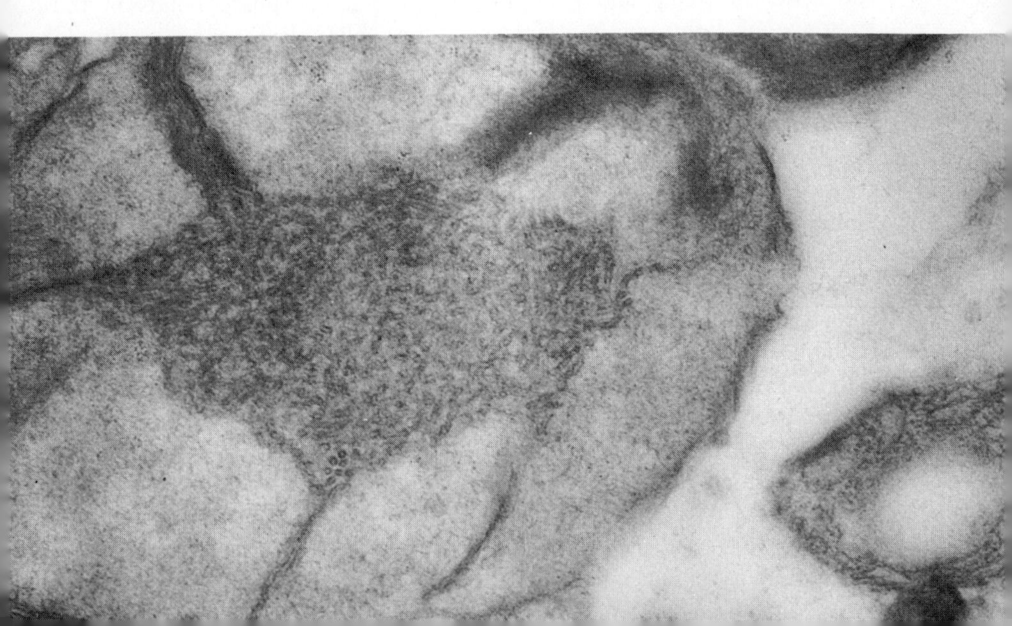

have never been exposed to light, and which thus possess only proplastids, nutrition is of the heterotrophic type, i.e., preformed food molecules must be absorbed and utilized by the cell. It is only when the proplastid has developed into the chloroplast that the autotrophic habit ("self-feeding" via photosynthesis) can be adopted.

Some biologists view the chloroplast as a sort of invading organism that by chance found its way into a previously non-green cell and converted it into the autotrophic type. In support of this hypothesis, they point out that if certain cells, such as the unicellular flagellate, *Euglena,* are grown at fairly high temperatures for several successive generations, the rest of the cell replicates faster than the chloroplast, giving rise to successively paler and paler green cells. Totally non-green cells may eventually be produced by the "dilution out" of the last chloroplast or proplastid. Such cells are permanently non-green and cannot regain the autotrophic habit. The disappearance of chloroplasts may also be induced by streptomycin treatment, or by treatment with other chemicals such as certain hydroxynitrobenzoic acids or aminotriazole. Thus the cell can be "cured" of the "invading" chloroplasts by high-temperature therapy or by chemical treatment.

Chloroplasts do not usually develop in root tissues, even those exposed to light, although in certain species, such as carrot and *Convolvulus,* root cells may "green-up" upon illumination. Why the plastids do not develop to maturity in the cytoplasm of some cells is unknown.

The intact chloroplast can be isolated from the fragmented cell by the technique of differential centrifugation, and can be shown to possess and retain for some time all the attributes of the photosynthetic apparatus of the cell. Isolated intact chloroplasts have been shown to fix C^{14}-labeled CO_2 and to incorporate it into sugars. They can also split the water molecule and generate energy-rich phosphate bonds in the presence of light energy. However, the chloroplast cannot maintain itself or reproduce outside the cell. If the chloroplast is really an "invader from without," it has become markedly dependent on the remainder of the plant cell for many aspects of its existence.

In addition to the chloroplast, the higher plant cell contains other types of plastids which lack the layered structure and photosynthetic apparatus of the chloroplast. These bodies include the *leucoplast,* which is colorless, and the *chromoplast,* which usually harbors a high concentration of bright yellow, orange, or red carotenoid pigment. Like the chloroplast, these bodies seem to be transmitted from one cell generation to the next by means of proplastid-like structures carried in the maternal cytoplasm. The leucoplast serves as a storage center for the cell's reserve food materials, such as starch grains (Fig. 15), and therefore probably possesses the enzymatic machinery necessary to synthesize such materials from smaller precursor molecules. The function of the chromoplasts

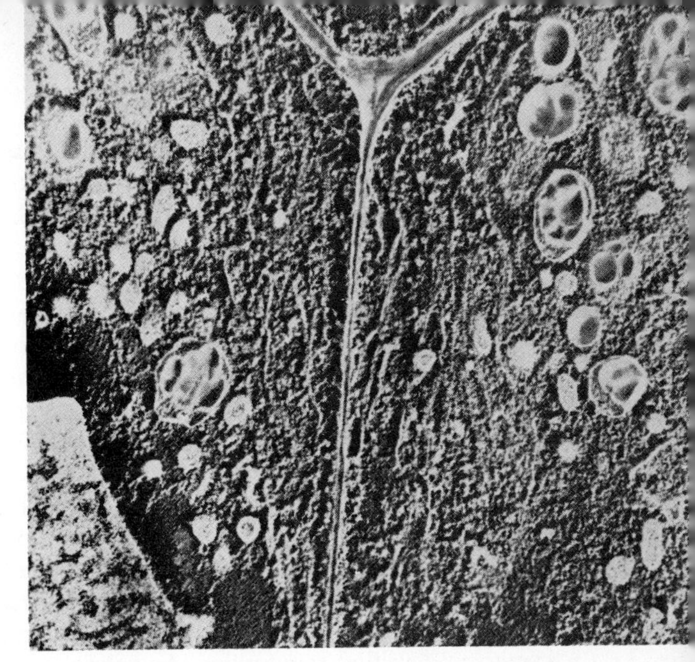

Fig. 15. An electron micrograph of a thin slice of plant tissue, showing the cell walls separating three cells, and numerous starch grains in leucoplasts. The large body at the lower left is the nucleus. (Courtesy R. W. G. Wyckoff, *The World of the Electron Microscope*. New Haven: Yale University Press, 1958.)

is obscure. They generally do not develop in the cell if chloroplasts are present. In the ripening process of a fruit such as a tomato, the green-white-red transition reflects three successive stages of development: the dominance of chloroplasts, the decline of chloroplasts, and the rise of the carotenoid-laden chromoplasts. The cause of these transitions is not understood.

The Cell Wall

Plant cells differ from all other cells in their possession of a fairly rigid outer case which resembles a box. Indeed, it is no great exaggeration to say that "the plant cell lives in a wooden box," for the chemical components of the cell wall include those that give wood its rigidity and high tensile strength. This wall is non-living and metabolically inert. It appears to be a secretion product of the protoplast of the cell, and is formed in successive layers as the cell goes through various stages of its developmental cycle (Fig. 16). The first layer formed after cell division is the *middle lamella*, which is at first composed largely of jelly-like pectin compounds, but is later infiltrated with tougher cellulose, polysaccharides of various types, and even woody lignin.

Fig. 16. The various layers in the plant cell wall. (From Carl Swanson, *The Cell*. Englewood Cliffs, N. J.: Prentice-Hall, 1960.)

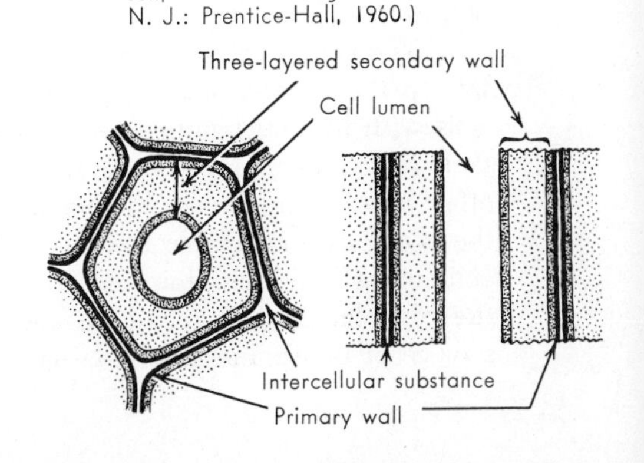

As the cell grows in volume, it must stretch the wall along with it. To do this, the contents of the cell exert great pressure against the wall and gradually stretch it to accommodate the growth. Some of this expansion is reversible (elastic) and some is irreversible (plastic). Since the cell wall itself usually thickens as the cell grows, it must acquire new material, either by incorporating new components into the framework of the existing wall (intussusception) or by adding a new layer alongside the older layers (apposition). By one means or another, the cell is always able to synthesize new cell wall materials rapidly enough to keep pace with its growth.

Once the cell has stopped expanding, rigid secondary material, which may have started entering the cell wall during the latter phases of growth, is added at a greatly accelerated rate. The new material is mainly lignin, cellulose, and other large molecules. Since the cell is no longer growing, the thickening of the wall decreases the volume of the protoplast. In some cells, such as fibers and tracheids, the wall may become so thick that it occupies almost the entire volume of the cell. If this happens, the cell dies and leaves its wall to serve in support or conduction.

The secondary wall is perforated by numerous *pits,* areas in which no additional wall deposition occurs on top of the middle lamella and primary wall. In mature, living cells, such as parenchyma cells, these pits are simple cylinders of open space that run from the innermost portion of the secondary wall to the outermost portion of the primary wall. In cells that are dead at maturity, such as tracheids, vessels, and fibers, the cylindrical area is covered by a flange of secondary wall material called a *border;* such pits are called *bordered pits* in contrast to the *simple pits* mentioned above (Fig. 17). The function of these pits, if any, is not definitely known. The cell wall, both secondary and primary, is also perforated by holes through which pass strands of cytoplasm, the *plasmodesmata.* These strands bind the protoplasts of neighboring cells together into one large community, the *symplast.* These connections facilitate passage of materials from one cell to another by providing a continuous cytoplasmic pathway free of differentially permeable membranes.

The wall, because of its rigidity, gives a certain form and minimum size to the plant cell and thus serves as the skeleton of the plant. In cells with heavily lignified secondary walls, such as those in wood, this skeleton can maintain the size and shape of the cell in the absence of any other supporting forces. In thin-walled cells, such as those in the leaf, the walls are too pliable to hold their shape without the support of the contents of the cell. Lacking this support, which comes mainly from the vacuole, the cell goes limp and shrinks in volume; to the eye, such a piece of tissue seems wilted.

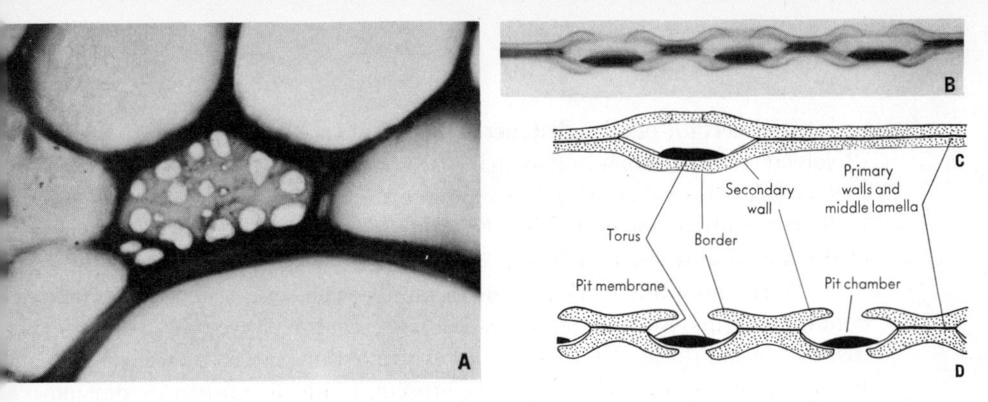

Fig. 17. Simple and bordered pits. (A) Simple pits in tobacco parenchyma. (B, C, D) Bordered pits in wood of hemlock. (From K. Esau, *Plant Anatomy*. New York: John Wiley & Sons, Inc., 1960.)

The Vacuole and Membranes

What is the vacuole, and how can it keep a cell fully expanded? The vacuole appears to be a "tank" in the cytoplasm, surrounded by a differentially permeable membrane and containing a watery solution of salts, organic molecules, and waste products of the cell's metabolism. When a cell is young, it possesses numerous small vacuoles that comprise only a small percentage of the total volume of the cell (Fig. 18). As the cell grows, the vacuoles gain in size and ultimately coalesce into one large central vacuole. This vacuole, which may occupy 90 per cent or more of the total volume of the cell, is surrounded by a membrane (the *tonoplast*) that is also the inner membrane of the cytoplasm. The tonoplast, composed of protein and fatty phases, has the property of *differential permeability;* that is, it permits the rapid passage of molecules of

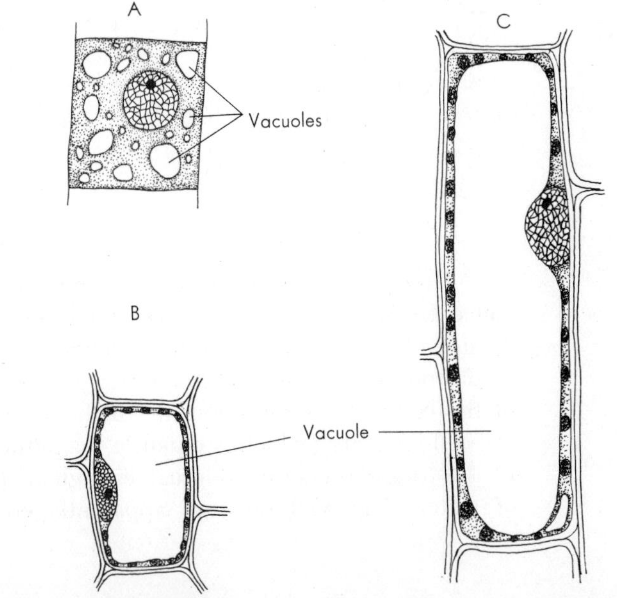

Fig. 18. Change in vacuolar system during growth of a cell. (A) Newly produced cell; (B) partly expanded cell; (C) fully expanded cell.

water and certain other substances, but allows a slower passage of most dissolved solutes, some of which it practically bars altogether. This fact enables the vacuole to exert a force on the cytoplasm and the cell wall; the motive power behind this force is merely the kinetic energy of the diffusing water molecules.

To see how this works, let us examine a cell with a central vacuole containing salts, sugars, amino acids, etc., that is sitting in a tank of distilled water (Fig. 19). As we know from the kinetic theory of matter, the molecules of all substances are constantly in rapid random motion, their average velocity being determined by (and actually a measure of) the temperature. Since water moves through the membranes of the cell much more rapidly than does any other substance, we may, for simplicity, consider only the movement of the water molecules. Since the cell's vacuole contains solutes in appreciable quantity, its water molecules are relatively more dilute than are those of the pure water outside the cell. This means that, in any given area of the membrane (for our present purpose, the outer and inner cytoplasmic membranes are equivalent), more water molecules will enter than will leave in any given time. This rapid, unequal two-way diffusion of water molecules through the membrane increases the volume of the vacuole and causes it to press the cytoplasm against the cell wall. Such pressure is referred to as *turgor pressure,* and such a cell is said to be *turgid.*

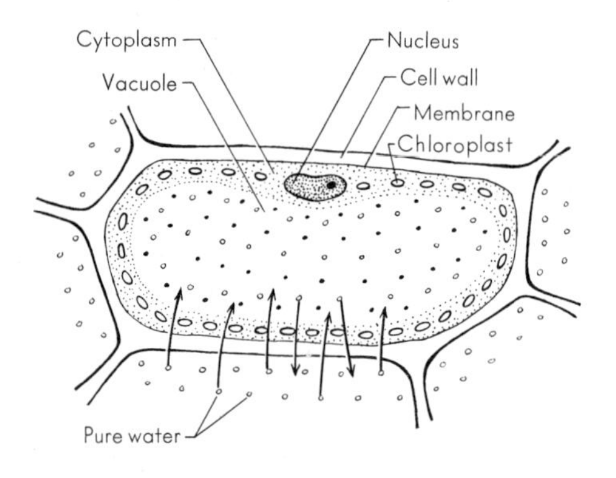

Fig. 19. The generation of an internal pressure against the cell wall as the result of differential rates of diffusion of water into and out of the cell. The vacuolar solute molecules (black dots) "dilute" the water, making their outward diffusion pressure lower than that of the incoming molecules from pure water.

How long does the water continue to enter? Theoretically, if the concentration differences of water molecules between the inside and outside of the cell were the only determinant, the net movement inward would never stop. Actually, however, it ceases when the turgor pressure of the cell reaches the point at which the back-pressure of the wall on the cell contents (which is equal in magnitude and opposite in direction to the turgor pressure) is great enough to prevent a further net entry of water. The wall pressure apparently compresses the water in the vacuole, thus lessening the difference in water concentration between the inside and outside of the cell, and it also activates the individual water molecules inside the vacuoles, thereby increasing their rate of bombardment of the membrane. A dynamic equilibrium is established, and net water entry ceases, although water molecules continue to move rapidly through the membrane in both directions. This diffusion of water through a differentially permeable membrane is known as *osmosis,* and the pressure developed by the process is called *osmotic pressure,* which is equivalent to *turgor pressure* and equal and opposite in direction to *wall pressure.*

It should now be clear that if an osmotic system such as the cell is placed in a solution in which the concentration (or more properly, the diffusion pressure) of water molecules is lower than that of the vacuole, the inward and outward diffusion of water will result in a net *loss* of water from the vacuole. Eventually, the cell will lose its turgor pressure and become flaccid and limp. If this process is continued, the vacuole volume will shrink so much that the cytoplasm will withdraw from the cell wall, producing a condition known as *plasmolysis* (Fig. 20). The external solute concentration that produces incipient plasmolysis is indeed a measure of the internal osmotic concentration. Although old and somewhat primitive, the plasmolytic method of determining osmotic concentration of a cellular vacuole is still probably the best we have. Most other methods require expressing the sap from cells, which is likely to alter the vacuolar contents.

In the plasmolyzed cell, the space between the outer cell membrane and the wall is occupied by the plasmolyzing solution, although the cytoplasm may continue to be in contact with the cell wall in the regions where plasmodesmata pass from cell to cell. Sometimes plasmolysis is so extreme that these plasmodesmata are also severed, and the entire protoplast contracts into the middle of the cell. Even such severely plasmolyzed cells, however, quickly recover when replaced in a solution in which the concentration (or diffusion pressure) of water molecules exceeds that of the vacuolar contents. Successive plasmolysis and deplasmolysis does not permanently injure the cell and, indeed, must occur almost daily in the leaves of some

Fig. 20. Plasmolysis of cells of red onion bulb scale. (From K. Esau, *Plant Anatomy*. New York: John Wiley & Sons, Inc., 1960.)

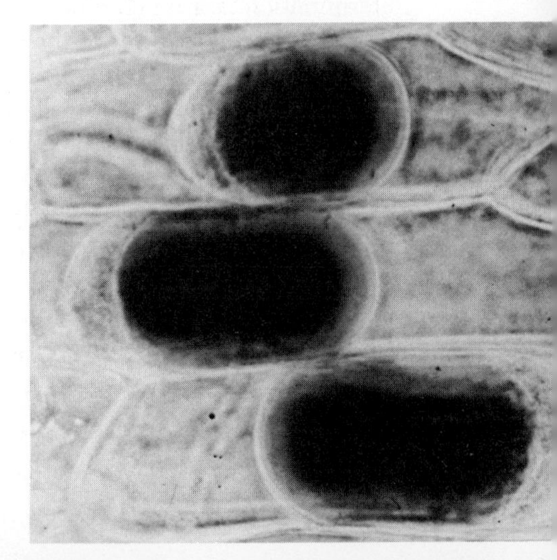

plants growing under conditions of water deprivation.

The direction of net water movement is not determined solely by the osmotic concentration of the vacuole. For example, a cell with a vacuolar solution equivalent to 0.3 M that is permitted to attain maximum turgor in distilled water will be unable to absorb additional water from any solution or any cell. In fact, it will lose water to a cell of lower osmotic concentration (for example, 0.1 M) if that cell is at incipient plasmolysis and thus subjected to no turgor pressure. The net water-accumulating tendency of the cell may be expressed as the difference (in some convenient units, such as molarity) between osmotic concentration (OC) and turgor pressure (TP). This quantity is sometimes referred to as *diffusion pressure deficit* (DPD) and indicates the extent to which the cell's potential water uptake has been realized. Thus:

$$\text{DPD} = \text{OC} - \text{TP}$$

In the first cell above, $\text{DPD} = 0.3 - 0.3 = 0$
In the second cell, $\text{DPD} = 0.1 - 0 = 0.1$

Since the second cell has the higher DPD, it would gain water from the first cell if the two were brought into contact, despite the higher osmotic concentration of the first cell.

A knowledge of the DPD of a cell can be used to predict whether it will gain or lose water from any solution. Experimentally, the DPD can be determined by immersing cells or bits of tissue in an osmotically graded series of solutions of some solute, such as sucrose. That solution in which the cell neither gains nor loses weight or volume represents its DPD. A simple immersion technique can be used, therefore, to determine the OC (concentration of external solution just causing incipient plasmolysis) and the DPD (concentration of external solution causing no change in weight or volume of the immersed cell). The TP can then be most conveniently calculated by subtracting DPD from OC, as its direct determination is a much more complicated matter.

The special features of the plant cell are thus roughly threefold: (1) The chloroplast enables the cell to convert radiant energy into chemical energy, thus rendering the cell autotrophic. (2) The cell wall serves as the skeleton of the plant body, encases each protoplast in a rigid framework, and allows local turgor pressures of great magnitude to build up. (3) The central vacuole, which occupies 90 per cent or more of the plant cell, not only collects waste products of plant metabolism, but also facilitates the uptake of water. This water is absorbed without the expenditure of energy by the plant, since the diffusion pressure of the water molecules themselves causes their net movement into the vacuole.

Cyclosis

With so many discrete organelles, each bounded by a differentially permeable membrane and separated from neighboring organelles by relatively large distances, how does the cell guarantee the necessary interchange of materials between the organelles? Part of the answer lies in the fact that the contents of many plant cells are in a state of fairly rapid movement, known technically as *cyclosis*. The entire cytoplasm rotates, either clockwise or counterclockwise, around the inner surface of the cell wall, carrying the various organelles along with it. In certain instances, as in the stamen hairs of *Tradescantia,* actively streaming cytoplasmic strands may extend through the vacuole. In some green cells, the chloroplasts can move independently and orient their broad surfaces either parallel with or perpendicular to the surface of the leaf, usually in response to altered light intensity. The mechanism of cyclosis is not well understood, but it seems tied to the cell's respiratory activity.

Cellular Types

The plant is a community of many different cell types that have all arisen from actively dividing cells at the apex and have developed through the process of *differentiation*. This is the process by which cells that appear to be similar assume specific and distinct morphological patterns. Differentiation is one of the major puzzles in biology. If, as we believe, each cell of the multicellular organism has arisen by division of the original $2n$ cell or zygote, then each cell should have an exactly equal complement of genes. As we have seen, this is not always true, since different cells may have different numbers of chromosomes as a result of certain aberrations in the mitotic process. But, as far as we know, there is no *qualitative* difference in the gene complements of cells that ultimately assume vastly different forms. A newly formed cell, then, has certain broad potentials and may develop along any of several morphological and physiological lines. Some unknown extranuclear or extracellular influence determines which line is followed. Once a cell has differentiated along a specific pathway, it ordinarily cannot revert to the undifferentiated state or assume a new form; when it enters one pathway of specialized development, it cannot adopt other pathways.

Physiologically, the plant consists of these types of cells (Fig. 21):

Meristematic cells: they retain their ability to divide repeatedly and give rise to the new cells of the plant body. Such cells are approximately cubical in shape, are small, thin-walled, and multivacuolate, and their nucleus is quite large relative to the rest of the cell. An exception is the *cambium* (p. 59). Its elongate cells possess a large central vacuole and maintain meristematic activity.

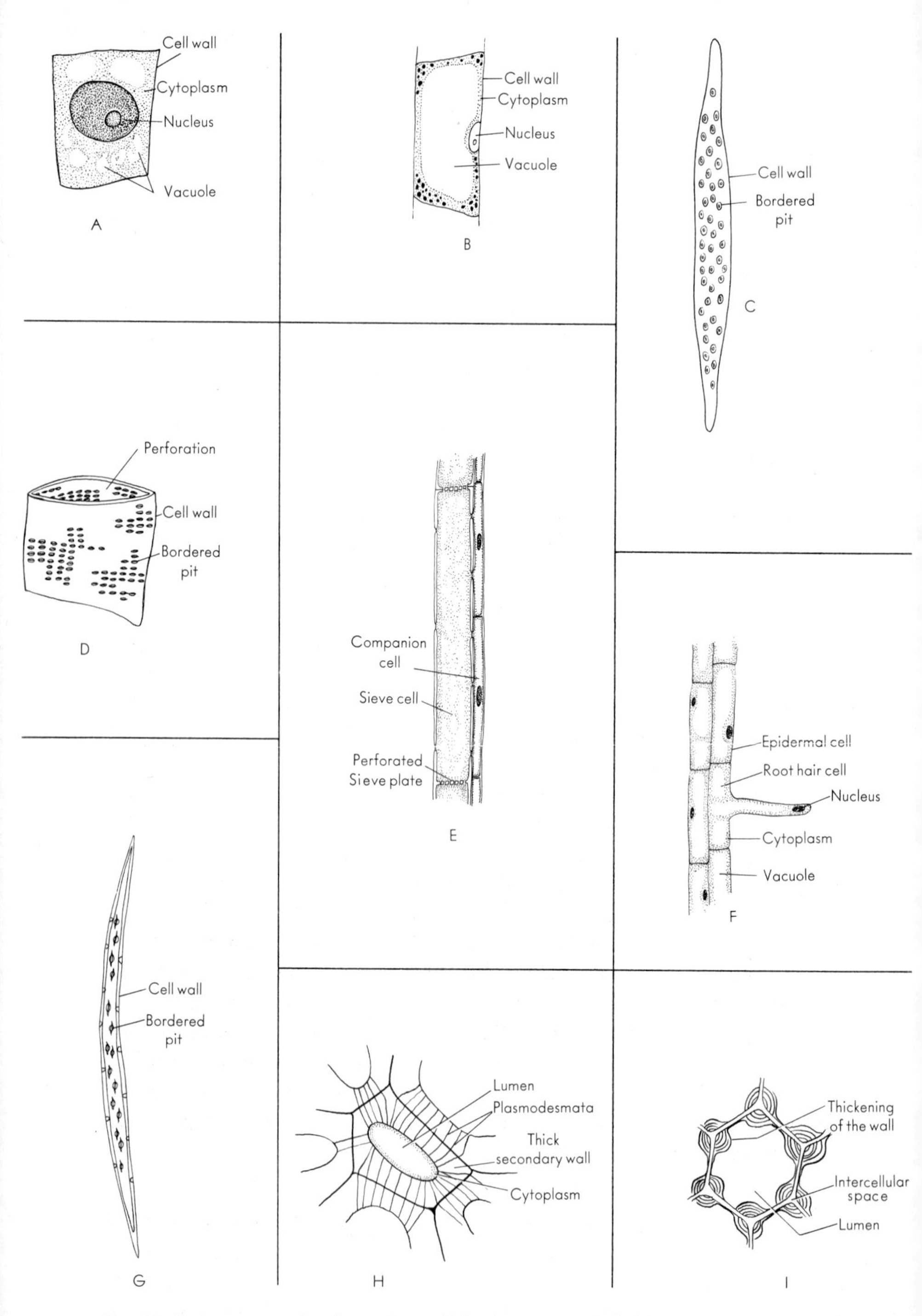

Fig. 21. Various types of cells in plants. (A) Meristematic cell. (B) Parenchyma cell. (C) Tracheid. (D) Vessel cell. (E) Sieve cell and companion cells. (F) Epidermal cell and root hair cell. (G) Fiber. (H) Stone cell. (I) Collenchyma.

Parenchymatous cells: these large, usually thin-walled polyhedrons keep their living contents and contain numerous simple pits and plasmodesmata. These relatively undifferentiated cells form the bulk of the primary tissues of the plant. They may contain chloroplasts, as in the leaf mesophyll, or may be adapted for the storage of water and reserve foods such as starch.

Supporting and conducting cells: they are all elongate but are otherwise greatly different from one another. One group, the heavily lignified cells, includes *tracheids, vessels,* and *fibers* that are all devoid of living contents at maturity. Vessels and tracheids possess a relatively large central space, the *lumen,* through which water is conducted. These lumina are sometimes blocked by inward extensions, called *tyloses,* of neighboring parenchyma cells and can no longer transport water. Fibers resemble tracheids, except that their wall is thicker and their lumen smaller; their major function is physically to support the organ. All these cells contain bordered pits, and the vessels have large perforations in their end walls. Another type of supporting cell, *collenchyma,* is thickened only at the corners of the cell, but otherwise resembles parenchyma. *Sieve tubes,* the main conducting elements of the phloem, are strings of elongate cells with perforated end plates, through which extend cytoplasmic strands that are continuous with the contents of the tubes themselves. Such sieve cells are devoid of nuclei at maturity; their nuclei are believed to reside in the small *companion cells* that generally lie alongside the sieve cells. Sieve cells may also become blocked and nonfunctional when the carbohydrate, *callose,* is deposited across the sieve plates at the end of the cell.

Protective cells: these generally have flattened surfaces that are often covered by a layer of a water-impervious substance, such as cutin or suberin, that they secrete themselves. In this category are the *epidermis,* a primary tissue, and the *periderm,* a secondary tissue derived from the cork cambium. The epidermis contains, in addition, such specialized structures as stomata, hairs, and glands.

Reproductive cells: they arise at particular times in the history of the plant, usually in response to specific environmental stimuli such as temperature and day length. The megaspore- and microspore-mother cells, which are diploid, give rise through their meiotic divisions to the mega- and microspores, which subsequently divide to produce the gametes, whose fusion completes the sexual cycle of the plant. Although other specialized cells, such as endodermis, latex vessels, glands, and idioblasts, exist in various plant parts, they need not be discussed here.

The cell, then, the fundamental functional unit of biology, is a very complex structure that consists of interrelated and interdependent organelles. It exists in a dynamic state, and its various parts are renewed by the ceaseless synthetic activity without which any cell would die.

Certain cells possess, in addition, the remarkable ability to synthesize complete replicas of themselves, replicas that in turn show a renewed ability to grow and synthesize. Others go through a profound metamorphosis and assume a highly specialized structure and function. Despite the tremendous recent advances in the study of the cell and its component organelles, biologists are still very far from understanding in any great detail the functioning of this complex structure.

3 Plant Nutrition

The green plant, like all organisms, requires three major classes of nutrients: (1) oxidizable organic foodstuffs, (2) mineral elements, and (3) water. The green plant differs from most organisms in being autotrophic with regard to its organic nutrition; that is, it has the ability to reduce atmospheric carbon dioxide to the level of sugar via radiant energy and the photosynthetic apparatus of the chloroplast, and is thus independent of any external supply of oxidizable fuel. In this chapter, we shall inquire into the mechanisms employed by the plant in satisfying its various nutritional needs.

Photosynthesis

When light of appropriate wavelengths is absorbed by the chloroplast, carbon dioxide is reduced to the level of sugar, and gaseous oxygen, equal in volume to the CO_2 reduced, is liberated. The direction of these changes is exactly the reverse of those accomplished during the oxidation of foodstuffs in the process of respiration, and, indeed, plants are important in the balance of nature because they restore to the air the O_2 needed for respiration by most organisms. Using the formula $[CH_2O]$ to designate the basic unit of a carbohydrate molecule (six of these units would yield $C_6H_{12}O_6$, or glucose), we can write the equation for photosynthesis as:

$$\underset{\text{Carbon dioxide}}{CO_2} + \underset{\text{Water}}{H_2O} \xrightarrow{\text{Light energy}} \underset{\text{Carbohydrate}}{[CH_2O]} + \underset{\text{Oxygen}}{O_2}$$

This equation, although properly balanced, gives an erroneous impression of the mechanism by which the reaction is accomplished. By the use of H_2O and CO_2 labeled with isotopic oxygen (O^{18}), biochemists have been able to demonstrate that the oxygen released in photosynthesis comes not from CO_2, but from water. In fact, the photolysis of water appears to be one key to the entire process of photosynthesis, for it represents a point at which light energy is made to do chemical work. Since there are two atoms of oxygen produced in the above equation, and each water molecule contains only one atom of oxygen, at least two water molecules must enter into the reaction to satisfy this requirement. To write a balanced equation truly representing the mechanism of the over-all reaction, we must add one molecule of water to each side of the equation:

$$CO_2 + 2H_2\underline{O} \xrightarrow{\text{Light energy}} [CH_2O] + \underline{O}_2 + H_2O$$

The oxygen evolved is derived from water entering the reaction (as shown by the underline), and the water molecule formed is different from either of the two split in the photolytic process. Here is a scheme that may help you visualize the basic aspects of the reaction:

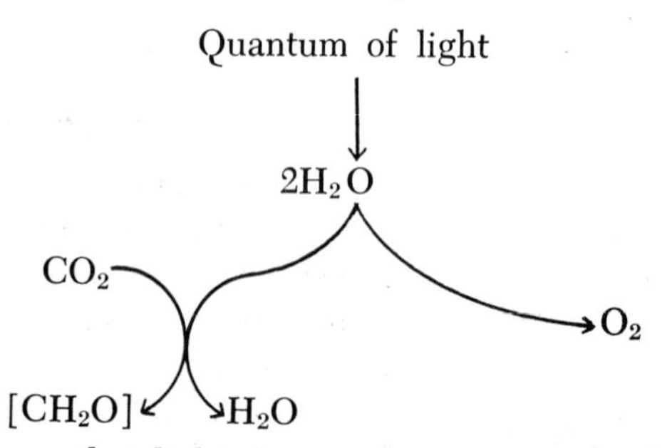

This scheme shows that light cleaves the water molecule, releasing oxygen, and that the "hydrogen" (or reducing power) also produced in the photolysis reaction is used in two ways: (1) to reduce CO_2 to $[CH_2O]$, and (2) to produce a new water molecule. This is clearly the briefest kind of shorthand, and there are many steps, some known and some unknown, that intervene in each of the simple reactions shown.

LOGISTICS OF PHOTOSYNTHESIS

The CO_2 entering into the reaction comes to the green cell of leaf or stem from the air by way of certain open pores, called *stomata* (singular *stoma*), in the surface of the leaf, and a much-branched system of interconnecting air canals. The leaf (Fig. 22), by far the main photosynthetic organ of the plant, can be visualized as several layers of actively photosynthesizing cells (the *mesophyll*) surrounded by a

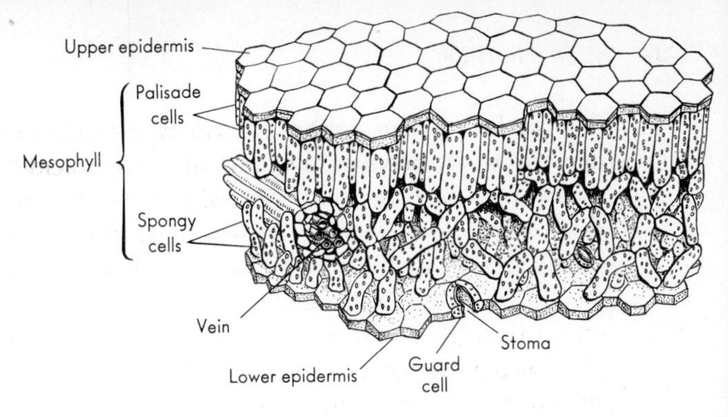

Fig. 22. A perspective view of the leaf. Note *epidermis* with *stomata*, *mesophyll* with numerous chloroplasts in each cell, and *veins* with xylem and phloem conducting cells.

protective layer (the *epidermis*) and supplied with conducting elements (the *veins*) that are equipped for two-way transport; they carry raw materials to the leaf and photosynthetic and other products away from the leaf. Such veins branch so profusely that no mesophyll cell is more than 1–2 cell diameters from contact with a vein.

The stomata that perforate the epidermis have the ability to open and close, in response to the turgor pressure of two sausage-shaped guard cells which surround them (Fig. 23). The inner walls of these *guard cells* are much thicker than the outer ones; thus when a guard cell is under great turgor pressure, the weaker outer wall will balloon out and carry the inner stronger wall along with it, causing the stomatal pore to open. When the guard cell is flaccid (perhaps owing to rapid water loss from the leaf on a hot day), the thick, elastic inner wall pulls the rest of the cell inward toward the pore area, effectively closing it. This control of the stomatal opening helps the plant prevent excessive water-vapor loss through stomata from an already partly wilted plant. Stomatal closure, however, also results in an inhibition of CO_2 absorption from the air, thus interfering with photosynthesis. Once the CO_2 molecule has passed the stomatal barrier, it enters a substomatal air chamber and connecting air passages, through which it can diffuse

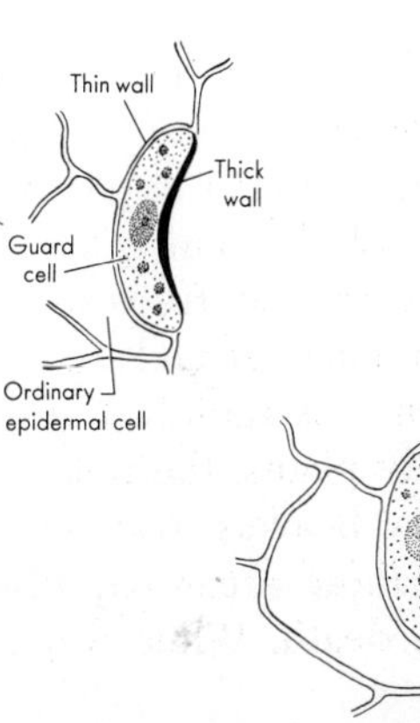

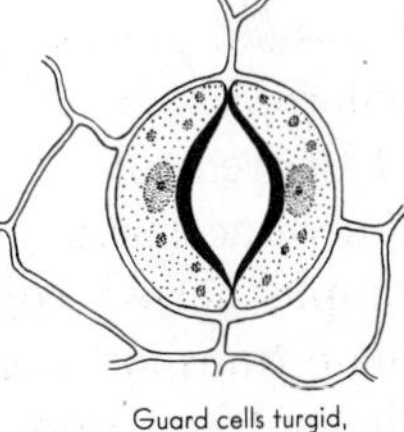

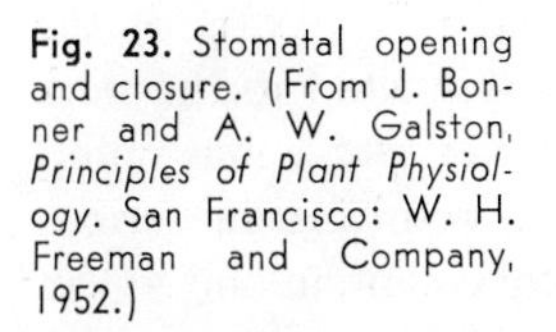

Fig. 23. Stomatal opening and closure. (From J. Bonner and A. W. Galston, *Principles of Plant Physiology*. San Francisco: W. H. Freeman and Company, 1952.)

throughout the leaf mesophyll. When it reaches the wet surface of a cell, it dissolves in water, becoming hydrated to carbonic acid (H_2CO_3), which is in turn partially neutralized by the cations of the cell to form bicarbonate ions (HCO_3^-). This bicarbonate represents a reservoir of potential CO_2 that can be tapped for photosynthetic purposes by the cell.

The oxygen released in photosynthesis is liberated to the outside world through the stomata, after passage from the surface of the mesophyll cell into one of the intercellular air spaces that joins up with the substomatal cavity. Closure of the stomata because of the loss of guard-cell turgor pressure would also prevent this gaseous exchange, and the oxygen would probably be re-absorbed in respiration, which would then produce CO_2 that could in turn serve as a substrate for photosynthesis. Closure of the stomata does not, therefore, prevent photosynthesis or respiration, but it does largely prevent net gas exchange with the outside world. Photosynthesis, which can proceed at 10–20 times the maximal rate of respiration, is limited, under conditions of stomatal closure, to approximate equality with respiration. It is clear, then, that stomatal closure that results from wilting causes a serious depression of the over-all photosynthetic activity of the plant.

The water required for photosynthesis and other activities of the leaf is absorbed from the soil by the roots of the plant through the purely osmotic activities described in the previous chapter. It then passes into the water-conducting tracheids and vessels of the xylem, located at the center of the root, and moves upward to the leaves through this tissue. The mechanism of this movement will be discussed later in this chapter. At the end of the vein, the water diffuses into neighboring mesophyll cells, passing from one cell to the other in response to DPD gradients. A certain amount of the water evaporates from the wet surfaces of the mesophyll cells and passes out of the stomata via the interconnecting intercellular air passages mentioned above.

The sugar and other organic materials produced in photosynthesis accumulate rapidly in the mesophyll cells. Some of the sugar is transformed into *starch,* a large polymeric molecule containing one to several thousand glucose units. The starch grains may be formed directly in the chloroplast or in the leucoplasts of nonphotosynthetic tissue. The bulk of the remainder of the sugar formed in photosynthesis is converted into the disaccharide, *sucrose,* which consists of one molecule each of the simple hexose sugars, glucose and fructose (Fig. 24). Sucrose is the main transport sugar of the plant. It finds its way into the sieve tubes of the phloem, through which it is rapidly transported to all parts of the plant. Since, in many woody plants, the phloem is restricted to the bark area, injury to the bark (by beavers, restraining wires, etc.) may interrupt the flow of sugar to those areas requiring it, resulting in serious growth inhibitions or even death. While water movement in the xylem

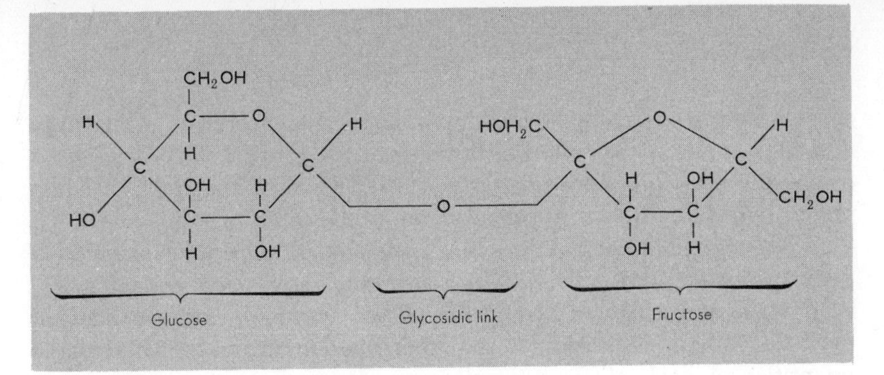

Fig. 24. The sucrose molecule, a disaccharide.

occurs in the dead, hollow tracheid and vessel cells, movement of sugars in the phloem is intimately connected with the vital activities of the sieve-tube cells in which they are conducted. An interruption of these activities may result in an undue accumulation of sugars and starch in the photosynthesizing mesophyll cells of the leaf, with an attendant decline in the rate of photosynthesis.

Thus, the maintenance of an optimal rate of photosynthesis in the leaf requires that adequate levels of light energy, water, and carbon dioxide be supplied to the leaf and that an adequate rate of transport of the products of photosynthesis be maintained from the leaf. With regard to light, almost all plants are "light saturated" at the intensity of bright sunlight, which approximates 10,000 foot-candles (Fig. 25). Individual leaves tend to be saturated at 1000 foot-candles or less, depending on the species, but because of the mutual shading of leaves on a plant, it takes several thousand foot-candles to saturate the entire plant. Some

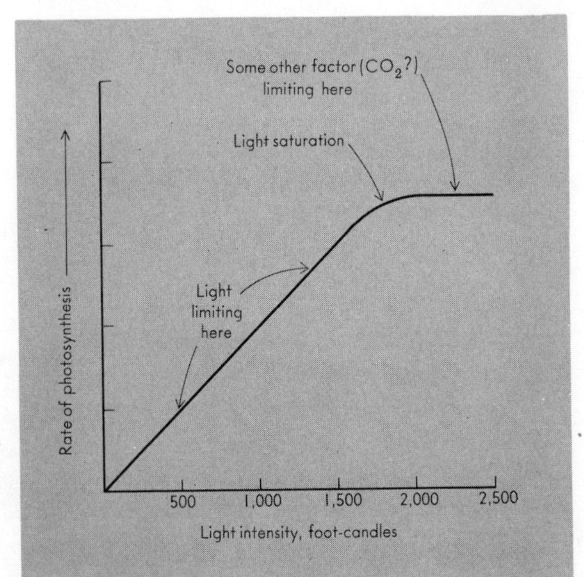

Fig. 25. A light-saturation curve for photosynthesis.

species, known as "shade-loving" types, can grow effectively in relatively dim light, such as that filtering through to the forest floor. These types compete successfully with other types ("sun-loving") that require much higher light intensities for optimal photosynthesis.

The carbon dioxide upon which we are all dependent exists as a trace gas in the atmosphere, comprising only three parts per ten thousand (0.03 per cent) of the air. This concentration varies somewhat over the face of the earth; it is higher over urban areas where large quantities of coal, oil, and gasoline are being burned, and lower in rural areas where massive photosynthesis is proceeding. The artificial elevation of the CO_2 content of the atmosphere raises the photosynthetic rate of plants that are well supplied with light and water, but it may also produce injurious effects in certain sensitive leaves.

Some students of evolution believe that the CO_2 content of the atmosphere may have varied considerably in recent geological times, and may have been responsible for certain changes of vegetation and climate. For example, an elevation in the CO_2 level would not only increase photosynthesis, and thus the amount of plant material on earth,

Fig. 26. The "greenhouse effect" of CO_2 in the earth's atmosphere.

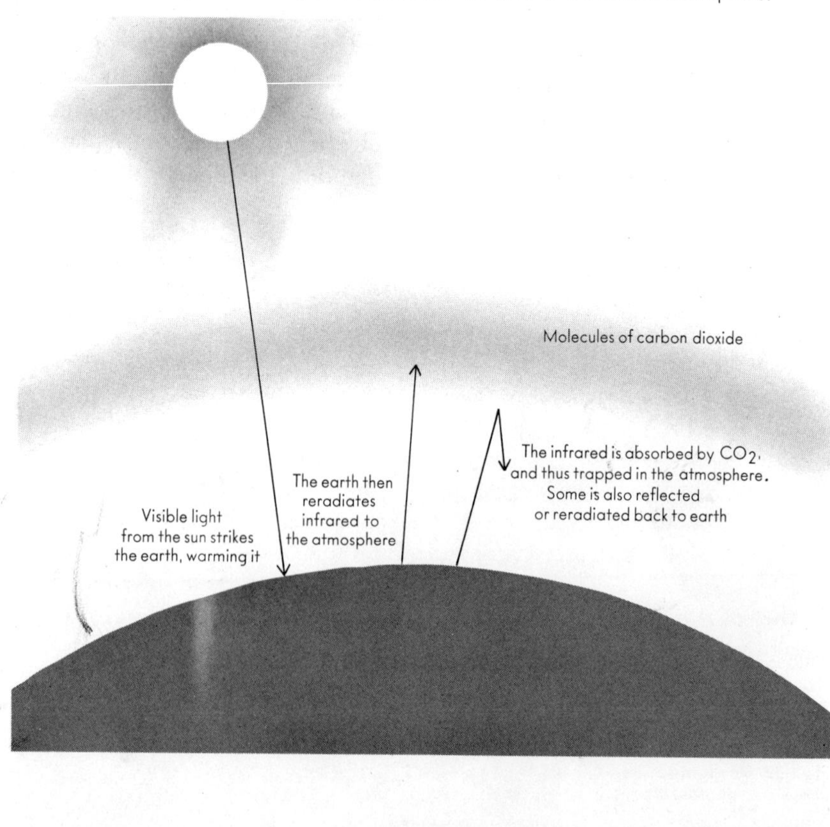

but would also cause a warming of the earth. This is true because the earth, heated by the sun, normally reradiates a portion of the absorbed energy back into space as infrared radiation. It happens that CO_2 absorbs specifically in this wavelength, thus preventing the complete escape of this heat energy and creating a sort of "greenhouse" over the earth (Fig. 26). Warming of the earth through such an effect could lead to partial melting of polar ice caps and glaciers, and to flooding of the low-lying land areas in which most of the major cities of the earth are located. Thus, our prodigious consumption of fossil fuels and the release of extra CO_2 into the atmosphere may have profound consequences for man. This process, however, tends to be self-limiting and perhaps cyclical, in that elevated temperatures and CO_2 levels will result eventually in elevated photosynthesis and a luxurious growth of plants, as in the Carboniferous era. This increase of CO_2 fixation in photosynthesis should ultimately lower the atmospheric CO_2 content significantly, resulting in the cooling of the earth and in a reversal of the cycle mentioned above.

As for water, the amount required in photosynthesis is only an infinitesimal fraction of that absorbed and evaporated by plants. For optimal photosynthesis, the leaf must be turgid and the stomata open; thus a suboptimal water supply depresses the photosynthetic rate, although only indirectly, since stomatal closure eventually inhibits the process through limitation of CO_2.

THE BIOCHEMISTRY OF PHOTOSYNTHESIS

The process of photosynthesis can best be approached by an analysis of three major questions: (1) How is light energy "captured" and made available for the performance of chemical work? (2) Through what pathway is carbon dioxide transformed into sugar? (3) Through what pathway is oxygen released from water? We shall examine each of these questions in turn.

Light, to be photochemically effective, must first be absorbed. Those molecules that absorb visible light are called *pigments*. The absorption of a quantum by a pigment is a function of the electron distribution pattern in the pigment molecule; the wavelength absorbed depends on such details as the number and position of double bonds in the molecule, the presence of aromatic rings, etc. In fact, the absorption of a quantum changes slightly the electron distribution pattern, and the altered form of the pigment is said to be "activated." Since we understand certain facts concerning the relation of wavelength absorbed to structure of the absorbing molecule, we can deduce the characteristics of a photoreceptor pigment in a particular photochemical reaction from the data relating wavelength to activity.

If various wavelengths of monochromatic light are permitted to fall on a green leaf and the photosynthetic rate is measured at each

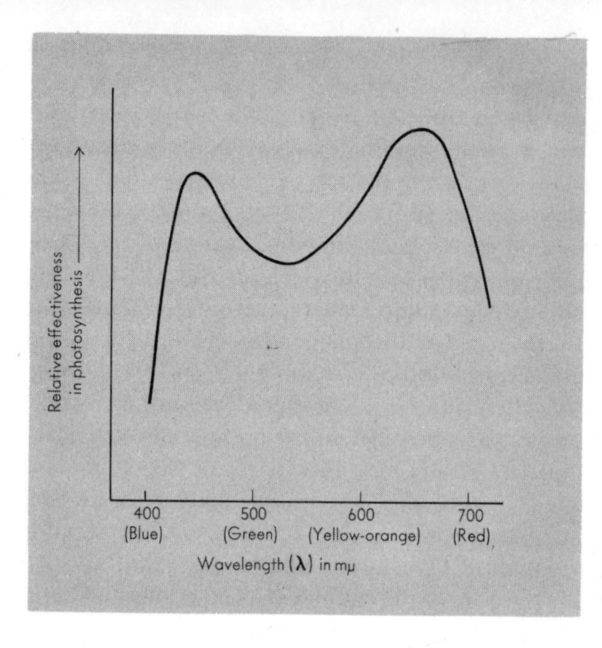

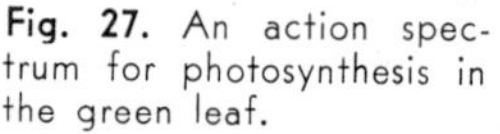
Fig. 27. An action spectrum for photosynthesis in the green leaf.

wavelength, blue light (near 420 mμ) and red light (near 670 mμ) are found to be maximally effective, and green light (c. 500–600 mμ) least effective (Fig. 27). This pattern can be understood in terms of the absorption characteristics of chlorophyll, the major pigment of the chloroplast. This substance, when extracted from the leaf, heavily absorbs exactly those wavelengths that are most effective in photosynthesis (Fig. 28).

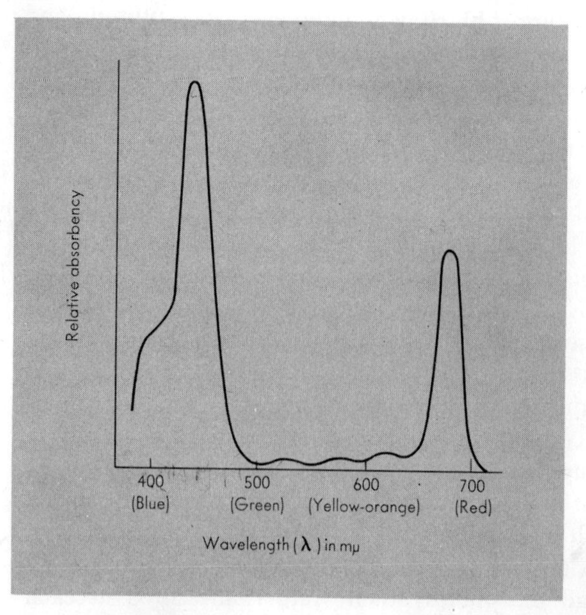

Fig. 28. The absorption spectrum of chlorophyll *a* in ether.

This coincidence between the "absorption spectrum" of chlorophyll and the "action spectrum" for photosynthesis is in fact the best proof we have that chlorophyll is the major receptor pigment in photosynthesis. From the details of the action spectrum, we may also infer that the yellow carotenoid pigments, also present in great quantities in the chloroplast, must be functional in light absorption for this process. Since carotenoids cannot operate photosynthetically in the absence of chlorophylls, it is generally assumed that photoactivated carotenoids pass their extra energy on to the chlorophylls, which ultimately perform the actual photosynthetic work.

In many algae that are red, blue-green, or brown in color, the action spectrum for photosynthesis differs markedly from that for green leaves (Fig. 29). The details of these complicated action spectra are well matched by pigments such as phycoerythrin (red), phycocyanin (blue), and others that are related to the bile pigments of our bodies. In the purple photosynthetic bacterium, the effective pigment is bacteriochlorophyll, a structural variant of chlorophyll, which, however, absorbs heavily in the green region of the spectrum and in the infrared, regions where higher green plants do not absorb heavily. In all these forms, the photosynthetic pigments are found in structurally complex bodies, either modified chloroplasts or smaller bodies called *chromatophores*. In all instances, the photochemical work performed involves the splitting of water or of some chemical analog of water, such as H_2S. The consummation of such a reaction apparently requires the high degree of structural organization found in chloroplasts or chromatophores.

Thus far we have noted the following energy transformations in photosynthesis: (1) The radiant energy of an absorbed quantum is

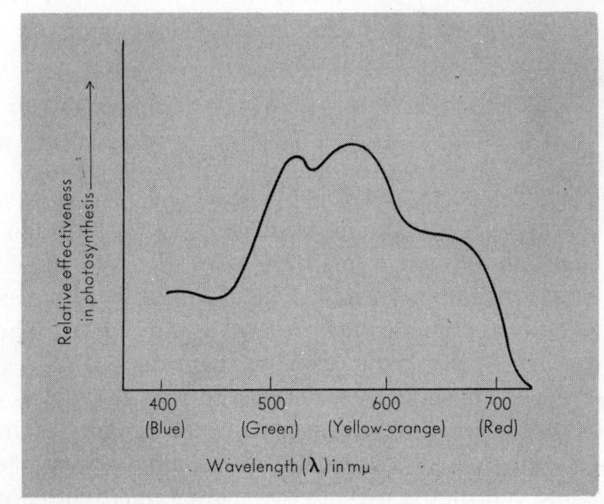

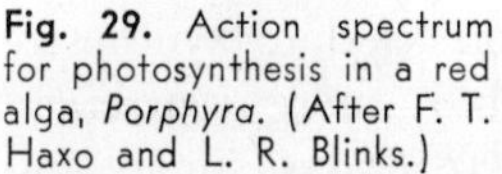
Fig. 29. Action spectrum for photosynthesis in a red alga, *Porphyra*. (After F. T. Haxo and L. R. Blinks.)

transformed into the energy of an activated pigment molecule:

$$\underset{\text{Pigment molecule}}{P} + \underset{\text{Quantum}}{h\nu} \longrightarrow \underset{\text{Activated pigment}}{P^*}$$

(2) The photoactivated pigment then cleaves the water molecule into smaller fragments, becoming "deactivated" in the process:

$$H_2O \rightarrow [H] + [OH] \quad (P^* \rightarrow P)$$

(3) The small fragments react in many ways, including the release of O_2:

$$2[OH] \longrightarrow O_2 + 2[H]$$

and the formation of reduced carrier molecules (RH_2):

$$R + 2[H] \longrightarrow RH_2$$

Another reaction of the split products of water is their recombination into the water molecule itself:

$$[H] + [OH] \longrightarrow H_2O$$

This reaction is, however, strongly exergonic (energy-releasing). The chloroplast is able to put this reaction to work, by causing it to synthesize an energy-rich molecule called adenosine triphosphate (ATP) from a precursor molecule (adenosine diphosphate, ADP) and inorganic phosphate (P_i):

$$[H] + [OH] \rightarrow H_2O \quad (ADP + P_i \rightarrow ATP)$$

(4) The energy of the ATP molecule can now be used to drive various chemical reactions, including the reduction of CO_2 to sugar by the reducing power generated in the light reaction:

$$CO_2 + 2RH_2 \rightarrow H_2O + [CH_2O] + 2R \quad (ATP \rightarrow ADP + P_i)$$

Thus the energy of the quantum has been converted to the chemical energy of the sugar molecule by passage through a photoactivated pigment, photolyzed water fragments, and ATP. The main function of light energy in photosynthesis, therefore, is to produce ATP through a complex of reactions called *photophosphorylation*. The subsequent reactions leading to the production of sugar from CO_2 can proceed entirely in darkness. It should be noted also that the sugar molecule itself can be oxidized, and the energy of this exergonic reaction used to synthesize additional ATP. While ATP is the readily negotiable "energy currency" of the cell that is used to drive numerous synthetic reactions, sugars and other stored foodstuffs represent a storehouse from which ATP can be produced by oxidation upon demand by the cell.

THE PATH OF CO_2 REDUCTION TO SUGAR

The pathway along which carbon dioxide is reduced to sugar has been elucidated in recent years by the use of radioactive carbon, C^{14}. This isotope decays with the emission of a beta-ray, which can be detected as a single ionization event in a Geiger-Müller-type counter. The biochemical pathway of the labeled carbon is then traced by the measurement of various chemical fractions isolated from a photosynthesizing cell at various periods after the introduction of the radioactive CO_2. Using this general technique, Calvin and his colleagues at the University of California found that the first stable product of photosynthesis in the unicellular alga *Chlorella* is a three-carbon phosphorylated compound called 3-phosphoglyceric acid (PGA):

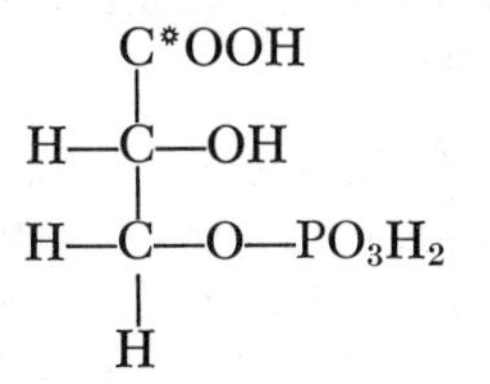

By stepwise degradation of this molecule, it can be shown that the terminal carbon of the carboxyl group is radioactive (shown by the asterisk) and must thus represent the altered form of the fed C^*O_2.

It might be anticipated that the fed C^*O_2 would couple with some two-carbon fragment to produce the three-carbon chain of PGA, but this seems not to be true. Instead, the C^*O_2 appears to attach to a five-carbon phosphorylated compound, *ribulose diphosphate,* to form an unstable six-carbon compound, which then spontaneously decomposes to two PGA molecules (Fig. 30). The ribulose diphosphate "acceptor" for CO_2 is resynthesized through the normal metabolic pathways in the cell. The phosphoglyceric acid formed from CO_2 is not yet at the reduction level of a carbohydrate, which corresponds to that of an aldehyde ($\mathrm{H{-}\overset{|}{C}{=}O}$) group, but is rather at the next higher oxidation step of a carboxyl group ($\mathrm{HO{-}\overset{|}{C}{=}O}$). The reduction to the aldehyde level is accomplished by means of the reducing power generated by the photolysis of water. This reducing power, which we have schematically indicated above as RH_2, is actually a reduced pyridine nucleotide, one of the normal oxidation-reduction carriers involved in electron

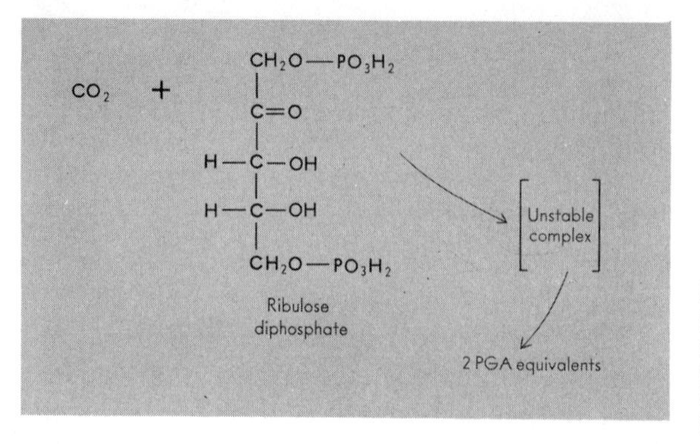

Fig. 30. Ribulose diphosphate, a 5-carbon phosphorylated sugar, combines with CO_2 and decomposes into two 3-carbon compounds equivalent to PGA.

transport in the cell. The final step in the production of a sugar from CO_2 via PGA, therefore, is the following:

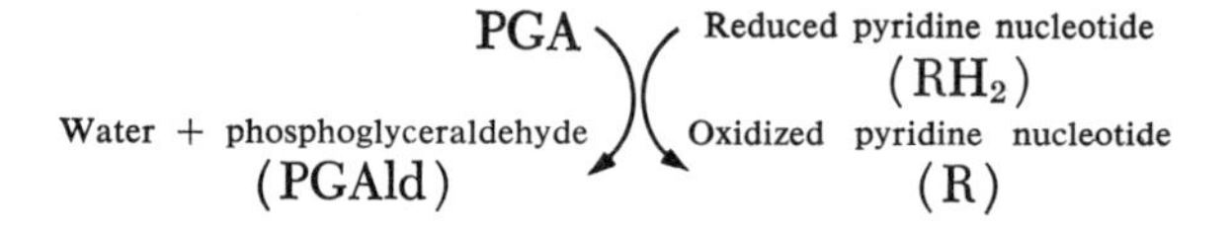

The phosphoglyceraldehyde, a sugar phosphate, has only three carbon atoms, while the smallest sugar accumulated to any extent has six carbon atoms. To produce this hexose, two PGAld molecules (or simple derivatives thereof) are made to combine "head to head" to yield a phosphorylated hexose, which can then be dephosphorylated to form the free hexose sugar, as in the following schematic sequence:

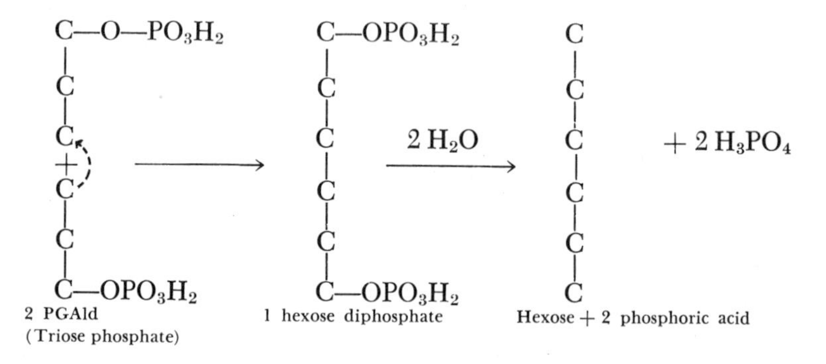

This hexose may then be easily converted to other hexoses, to disaccharides such as sucrose, or to giant polymeric molecules such as starch or cellulose that are built of thousands of hexose units strung together into long branched chains. The sugars may also be transformed to organic acids by oxidation, to fatty acids by reduction, and to amino acids by a combination of oxidation and ammonia uptake. Thus, the sugar produced from CO_2 in photosynthesis is the basic organic molecule used by the cells of higher plants for energy and for structural building blocks required by the cell.

OXYGEN EVOLUTION

The final question we have posed relative to the biochemistry of photosynthesis—what is the pathway of oxygen evolution?—is not answerable in precise terms at present. Most investigators feel that the first step involves the change of water into what are called free radicals of hydrogen [H] and hydroxyl [OH]. The "hydrogen" free radical would consist of a proton and an electron. Its charge would thus be zero, but its unfilled valence electron shell would give it high reactivity. Similarly, the "hydroxyl" free radical would have no net charge, but an unfilled electron shell, giving it high reactivity. The most probable pathway for oxygen evolution would appear to involve the fusion of 2[OH] radicals to a *peroxide,* which would then be decomposed with the release of oxygen. No known peroxide-decomposing systems in cells appear to be involved in this reaction, however, and its intimate details remain for future investigation to reveal.

Mineral Nutrition

In addition to the organic materials produced in photosynthesis, the plant requires a wide variety of mineral elements. These elements, absorbed from the soil by the root system and transported upward in both xylem and phloem, are required for structural purposes in the cell and for the formation of the specific catalytic molecules called enzymes, which regulate the metabolism of all cells. All these elements except one, nitrogen, are ultimately derived from the parent rock giving rise to the soil; nitrogen is ultimately derived from the atmosphere, mainly through the process of nitrogen fixation.

An essential element is one without which the plant cannot complete its life cycle. To discover which elements are required by a plant, a seedling is usually planted in highly washed pure quartz sand in a hard glass, glazed porcelain, or plastic container which is then supplied with a nutrient solution made up in carefully distilled water, and containing only the purest salts available (Fig. 31). Great care must be taken to exclude organic impurities, microbial contaminants, and dust, all of which may supply traces of mineral-element contaminants. In certain instances, especially with large-seeded plants such as the garden bean, it is necessary in addition to remove the cotyledons, for they may contain sufficient stored quantities of certain elements to eliminate the necessity of external sources. If the cotyledons are left on, it may take several generations of seed-to-seed growth in the purified nutrient media to demonstrate a requirement for particular elements.

When a test plant is grown under adequate photosynthetic conditions in a mineral nutrient medium containing all essential elements, it will develop vigorously and normally. From such an experiment, it is clear that the green plant is completely autotrophic for all organic molecules it requires, including vitamins, hormones, and miscellaneous complicated molecules. Why, then, are organic fertilizers ever required for plants? The answer lies not in the plant itself, but in the nature and structure of soil. Soil, derived originally from fragmented parent rock, is a highly dynamic and complex medium for plant growth. It includes (a) *rock particles* of various sizes, from coarse sand to finer silt to very fine clay particles, (b) *organic matter*, generally the remains of plant, animal, or microbial cells that have died and are now decomposing, (c) *living organisms* of various kinds, including bacteria, filamentous fungi, algae, protozoa, worms, insects, and even larger animals, (d) a *soil solution*, containing inorganic and organic materials in aqueous solution, generally as a thin film surrounding the rock particles, and (e) a *gas phase*, containing oxygen (which is required for root respiration and active absorption of minerals by root cells) nitrogen, carbon dioxide, and the trace gases of the atmosphere.

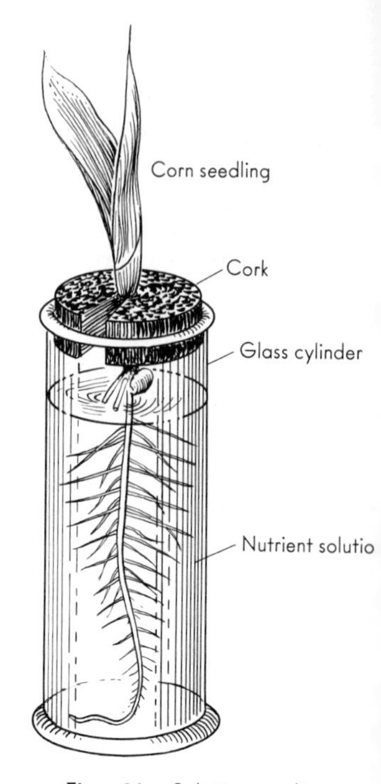

Fig. 31. Solution culture technique used by the plant physiologist, Julius Sachs, to investigate essentiality of minerals.

The vigorous growth of the plant in soil depends on the proper physical condition of the soil; if the soil particles are too closely packed together, there will be insufficient gas phase, and the absorption of materials by the aerobically respiring roots will decline because of a lack of oxygen. A soil is said to be in good "tilth" when it has good crumb structure, i.e., the fine soil particles are cemented together to form larger crumbs, which pack loosely to form a firm, yet well-aerated medium. It is this aspect of soil conditioning that is dependent on organic addenda, for the cementing of soil granules to form crumbs is accomplished by the soil microorganisms. These cells, feeding on the organic materials in the soil, produce the mucilaginous substances that result in a clumping together of soil particles to form crumbs. Thus, organic soil additives are required only if the physical condition of the soil demands it; under conditions of optimal growth in completely inorganic media, such as in well-aerated quartz sand watered with a mineral salt solution, organic additives produce no additional growth. There is also no evidence that organic fertilizers supplied to a soil will improve the quality of a plant for human or animal nutrition over and above that produced by growth in a well-balanced mineral solution. These facts should be remembered when one is confronted by spectacular claims of the proponents of "organic gardening."

If a plant is grown in a mineral solution lacking adequate quantities of some essential element, the plant will become unhealthy; it will develop typical deficiency symptoms for the element in short supply. A skilled botanist can learn to recognize the deficiency symptoms for each of the elements in a particular plant, and can ameliorate plant growth as it is proceeding by suitable additions to the soil or other medium in which the plant is grown. A more objective procedure is to harvest small bits of the growing plant from time to time and subject them to chemical analysis for the various elements. Any element found to be in short supply can then be added to the soil or solution, with prompt ameliorative effects on growth and vigor.

One of the first mineral nutrient solutions prepared for the growth of higher plants, by Knop in the 1880's, contained only three salts: calcium nitrate, potassium phosphate, and magnesium sulfate. In fact, these six major elements, together with the carbon, hydrogen, and oxygen assimilated in photosynthesis, are the ones required in largest quantities by higher plants. The three cations, K^+, Ca^{++}, and Mg^{++} may be combined in any way with the three anions, NO_3^-, $SO_4^=$, and $H_2PO_4^-$. In addition, most plants can use ammonium nitrogen (NH_4^+) or organic forms of N as well as or better than nitrate, and the phosphate may be supplied in any soluble form.

As commercially available mineral salts became purer and purer, it became clear that the basic three-salt solution of Knop was not complete. Many other elements are, in fact, required by plants, but in quantities much smaller than for the six major elements listed above. These additional elements are referred to as "micronutrient elements." They include iron (Fe^{++} or Fe^{+++}), manganese (Mn^{++}), Zinc (Zn^{++}), copper (Cu^+ or Cu^{++}), molybdenum (generally as MoO_4^-), boron (as $BO_3^{\equiv}$), and chloride (Cl^-). Some researchers also suspect that the plant may need very small quantities of yet other elements, such as cobalt (Co^{++}) or vanadium (generally as $VO_4^{\equiv}$), but absolute requirements for these elements have not yet been established. Certain specific plant cells have special requirements; for example, diatoms obviously need silicon to build their silica cell walls. As far as we know, this requirement is not general for plants.

In addition to those elements absolutely required for growth, plants can be shown to contain appreciable quantities of other elements. In certain instances, these inessential elements may actually stimulate growth or vigor. For example, wheat plants grown in the absence of silicon are markedly more susceptible to fungus attack than are similar plants grown in the presence of silicon, and beet plants grown in the presence of sodium (Na^+) produce larger and fleshier roots than those grown in its absence. Despite the beneficial physiological effects of such elements, they cannot, strictly speaking, be regarded as essential for the plant. Thus the plant is at present considered to need seventeen elements; four (C,H,O, and N) are derived ultimately from the atmosphere, while the remaining thirteen (K,Ca,Mg,P,N,S,Fe,Cu,Mn,Zn,Mo,B,Cl) are derived from the parent rock which gave rise to the soil.

NITROGEN FIXATION

The fixation of nitrogen is one of the most important, yet one of the least understood, reactions in all biochemistry. In the atmosphere, nitrogen is present as the very stable N_2 molecule. This molecule must somehow be destabilized and cleaved before reduction of the nitrogen to ammonia can occur. This complex of reactions is achieved by certain bacteria that live on the organic matter in the soil (for example, *Azotobacter,* an aerobic form, and *Clostridium,* an anaerobic form) and also by certain associations of bacteria living in swellings, or *nodules,* of particular plant roots (Fig. 32). These latter bacteria belong to the genus *Rhizobium,* and the host plant is usually a member of the family *Leguminosae,* although other host plants have recently been found to fix nitrogen in association with bacteria. In addition, certain blue-green algae (such as *Nostoc*) and photosynthetic bacteria (such as *Rhodospirillum*) are capable of fixing atmospheric nitrogen.

The association of two organisms for mutual benefit is called *symbiosis.* Since neither the *Rhizobium* nor the host plant alone can fix and reduce atmospheric nitrogen, the biological complex in the nodule must be regarded as a symbiotic association of bacterium and host. Recent investigations have revealed that the invading organism enters by way of curiously curved root hair cells. This deformation of the root hair is probably produced, in the first instance, by growth hormones of the auxin group (see Chapter 4) excreted by the bacterium. Once inside the host cell, the bacterium causes the formation of an *infection thread,* which runs from the tip of the root hair cell wall through the center of the cell, and probably contains all of the bacterial cells. The final result of this invasion is a prodigious overgrowth of the root cells to produce the warty protuberances called nodules.

These leguminous nodules frequently contain a red pigment called leghemoglobin, which is closely related to animal hemoglobin. It appears to be involved in nitrogen fixation, in that nodules lacking leghemoglobin are incapable of fixing N_2, while those containing the pigment are capable of fixation. The activation of nitrogen is brought about by some cellular catalyst capable of being inhibited by molecular hydrogen. The first stable detectable product of fixation is ammonia (NH_3), but the path of its production from nitrogen is as yet unknown. As this chapter is

Fig. 32. Nodules on soybean roots. (Courtesy The Nitragin Company.)

being written, news has come that researchers have succeeded in obtaining cell-free homogenates of *Clostridium* that are capable of fixing atmospheric nitrogen. The clarification of the chemistry of this process should now proceed more rapidly.

Most plants readily absorb and utilize nitrogen that is supplied in the form of nitrate (NO_3^-) even though the final form in which such nitrogen is incorporated into plant materials is highly reduced, such as the amino group ($—NH_2$). A recently discovered enzyme called *nitrate reductase* is able to reduce nitrate to ammonia, by means of reduced respiratory carriers such as pyridine nucleotides. This enzyme is believed to contain molybdenum at its active center, and this may be the only metabolic role for molybdenum, since plants supplied with NH_3 can grow without Mo, while those fed NO_3^- require Mo. The reduction of NO_3^- to NH_3 may proceed through such intermediates as hyponitrous acid (HONO) and hydroxylamine (NH_2OH).

The over-all cycle of nitrogen in nature thus involves passage between a free, gaseous form, comprising 80 per cent of the atmosphere, and a fixed form in the soil or biological system (Fig. 33). Fixation involves a transformation to ammonia, which may either be absorbed per se by plant roots, or may be first oxidized to nitrate by other soil microorganisms prior to absorption. In the cells of the plant, the nitrate is reduced back to ammonia, which is then coupled onto certain organic acids to form amino acids and ultimately proteins. These materials are ingested by animals and are transformed to animal proteins and nitrogenous waste products such as urea and uric acid. Eventually, all animals and plants die and are decomposed in the soil to simple nitrogenous materials, which may recycle again and again through biological systems. Also, the nitrogen may be returned to the atmosphere as free molecular nitrogen, through the actions of certain denitrifying bacteria on nitrate.

FUNCTIONS OF THE VARIOUS ELEMENTS IN THE PLANT

The major elements involved in photosynthesis, (C,H,O) plus nitrogen and phosphorus, constitute the main building blocks of the plant

Fig. 33. The nitrogen cycle in nature.

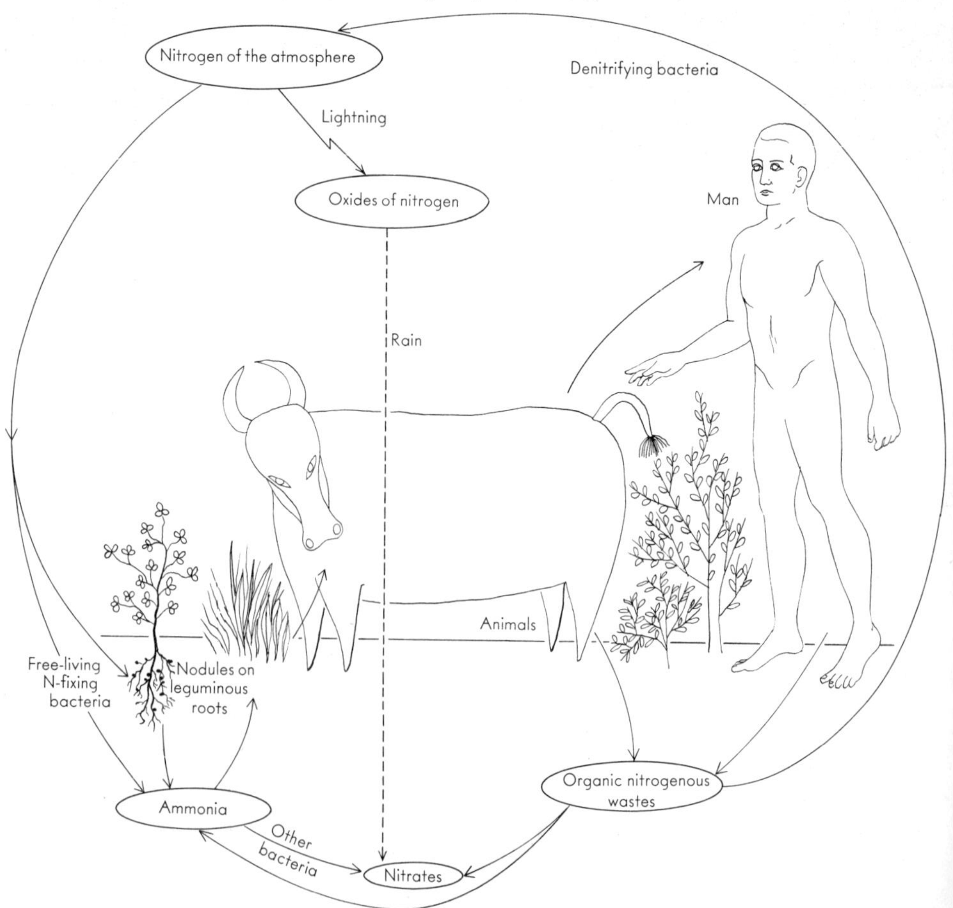

body. For example, the cell walls that comprise the plant skeleton are composed practically exclusively of C,H, and O; the proteins that comprise the major organic constituent of the cytoplasm are composed mainly of C,H,O, and N, and the nucleic acids that constitute a large bulk of the nucleus and some of the cytoplasm are made of C,H,O,N, and P. In addition, fats and carbohydrates, which may also constitute large percentages of the mass of the cytoplasm of particular plant cells, are built of C,H, and O.

Of the other twelve elements required by plants, another three are used mainly for structural purposes. *Sulfur* is a constituent of several amino acids that constitute the organic structural units ultimately built up into proteins. Thus even though relatively small quantities of sulfur are required by plant cells, almost all of it serves an important structural function, for without the sulfur-containing amino acids, many important proteins of the cell could not be synthesized. Sulfur is also present in such substances as glutathione and thioctic acid, which are widely distributed in plants and are believed to play a role in oxidation-reduction reactions because of their ability to change reversibly from the reduced form (–SH, or sulfhydryl) to the oxidized form (–S–S–, or disulfide).

Calcium serves a variety of functions, the major quantitative one being its incorporation into the structure of the middle lamella of the cell wall. It forms an insoluble salt when coupled with certain acidic components of the jelly-like pectins of the middle lamella. The introduction of calcium into the cell wall, therefore, serves to stiffen and make rigid a previously semifluid structure. *Magnesium,* a close chemical analog of calcium, is an essential part of the chlorophyll molecule. It occupies a central location in this molecule, being attached to each of the four pyrrole rings either by direct covalent bonds or by "secondary valences." In cases of magnesium deficiency, severe and typical chlorophyll deficiency (*chlorosis*) is manifested in the leaves. Magnesium is also known to be a specific cofactor for several enzymes present in all cells.

Phosphorus is present mainly as a structural component of the nucleic acids, DNA and RNA, and as part of certain fatty substances, the phospholipids, which are believed to play an essential role in the structure of the membrane. A deficiency of phosphorus is thus very serious for the cell, preventing, as it must, the formation of new genetic material in nucleus and cytoplasm and of new membranes around the surface of the cell and the various cell organelles. Phosphorus is also critically involved in all energy-transfer steps in the cell, since compounds such as ATP are composed of three phosphates coupled to a complicated ring structure. The two terminal phosphate groups in such a molecule are believed to be structurally different from the first phosphate group, for hydrolysis of the two terminal bonds yields much more energy than does hydrolysis of the first group. For this reason, the latter two bonds are called "energy-rich" phosphate bonds, and are usually symbolized by ~P to distinguish them from ordinary bonds, symbolized by —P. Thus ATP is generally written as A—P~P~P. The cleavage of this molecule into A—P~P + free phosphate releases relatively large amounts of energy which can be used to drive various energy-requiring reactions, such as the coupling of two amino acids to form a dipeptide. The ADP produced by the breakdown may be built back up into ATP by the energy released in oxidative reactions. Thus:

$$A{-}P{\sim}P + P + \text{energy} \rightleftharpoons A{-}P{\sim}P{\sim}P$$

Although phosphorus, magnesium, calcium, and sulfur have other roles to play in the cell besides the structural ones mentioned above, the structural role is quantitatively the most important for these elements.

The remaining eight elements (K,Fe,Mn,Cu,Zn,Mo,B,Cl) play their major roles as essential parts of catalytic entities in the cell. It is well known that many important enzymes consist of specific proteins to which are attached special chemical entities called *prosthetic groups* or coenzymes. These groups may consist entirely or in part of metallic elements such as Fe,Cu,Mn,Zn, or Mo. We have already mentioned that molybdenum appears to be involved in the functioning of the enzyme called *nitrate reductase,* which reduces nitrate to ammonia. Similarly, *copper* is a part of certain oxidative enzymes, such as tyrosinase and ascorbic oxidase, which oxidize, respectively, phenolic substances such as the amino acid tyrosine and vitamin C (ascorbic acid). *Iron* is a part of many important enzymes, including the respiratory electron carriers called *cytochromes* and the oxidative enzymes *peroxidase* and *catalase.* In all these enzymes, the iron is present in the prosthetic group as heme (an analog of chlorophyll), in which a central iron atom is connected to four pyrrole rings joined into a large cyclic structure. Iron functions in such enzymes by virtue of its reversible oxidation and reduction (Fe^{+++} + electron $\rightleftharpoons$ Fe^{++}). Copper probably functions in the same way (Cu^{++} + electron $\rightleftharpoons$ Cu^{+}), and it is likely that manganese and zinc play similar roles.

Potassium is known to activate several important enzymes, though it has never been isolated as part of an enzyme system. *Chloride* is known to stimulate photosynthetic phosphorylation, but its exact role in this process has never been exactly defined. *Boron*'s role is completely unknown. Boron deficiency generally results in the death of the meristematic cells, but here, too, the exact mechanism by which it acts is obscure. Because boron is known to form complexes with sugars and related molecules, it has been suggested that its function in the plant may involve sugar transport. In fact, experiments with labeled sugars have supported the view that boron may, under certain conditions, speed the rate of sugar movement in the plant.

Water Economy

Water is the most abundant component of active plant cells, constituting 90 per cent or even more of the fresh weight of some tissues. Even very dry and dormant seeds and spores have at least 15 per cent moisture. Plants continuously absorb water from the medium in which they grow, and evaporate water into the environment about them. The amount of water passing through a plant is thus much larger than the amount contained in it at any one time. In fact, the evaporation of water by the aerial parts of plants, a process called *transpiration,* can exert a major effect on the climate of an entire geographical region (Fig. 34).

Transpiration is an inevitable consequence of the architecture of the leaf and of other aerial parts of most plants. As we have seen, the leaf is essentially a layer of wet, photosynthetically active cells, well supplied with vascular elements and encased in a fairly waterproof but perforated layer, *the epidermis.* The wet cells of the mesophyll evaporate large

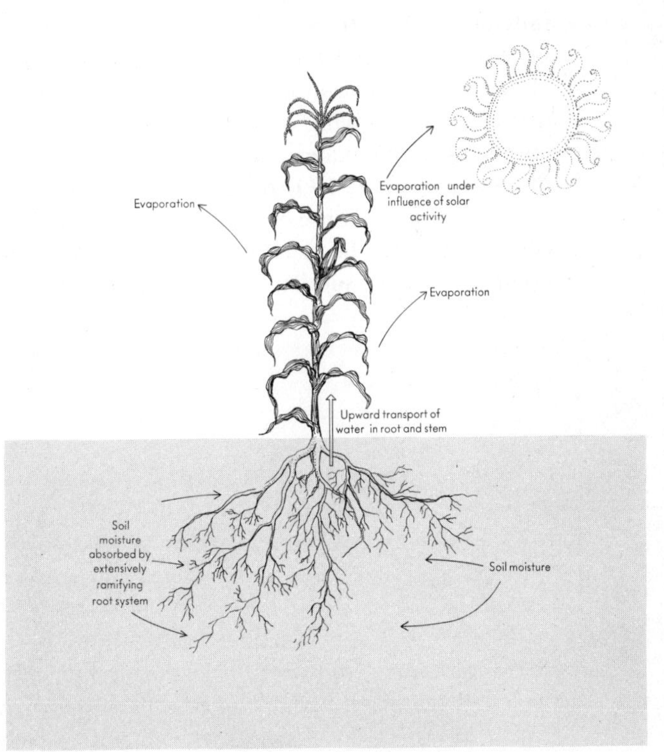

Fig. 34. The plant as a wick for the evaporation of soil moisture.

quantities of water into the intercellular air spaces. Because of the diffusion of the water vapor molecules, much of this water finds its way to the stomata, through which it passes to the outside world. Under conditions of high humidity and still air, the back diffusion of water vapor may greatly diminish net water loss from the leaf. Under conditions of lower humidity and turbulent air, the water vapor is quickly removed from the area surrounding the leaf, thus steepening the diffusion gradient from leaf to atmosphere and greatly increasing the rate of transpiration.

In most plants, epidermal cells are well covered with a waxy, water-impervious layer called the *cuticle*. Where such a thick waxy cuticle is present, water loss occurs almost exclusively through the stomata. If the stomata are closed, transpiration is greatly diminished or even completely halted. We have already discussed how the stomatal aperture is controlled by the turgor pressure of the guard cells; when the plant is deficient in water, the guard cells tend to become flaccid, thus closing the stomatal aperture and restricting water loss. Obviously, such a mechanism can serve to protect the plant against excessive desiccation.

Even when the plant is well supplied with water, the guard cells may be flaccid and the stomata closed. In recent years, it has been discovered that the CO_2 content of the substomatal cavity is a prime regulator of stomatal opening in many plants. If the CO_2 concentration falls below the 0.03 per cent normally present in the air, the guard cells become turgid and the stomata open. This condition is usually brought about by illumination of the guard cells, causing photosynthetic activity and an accompanying diminution of the CO_2 content of the surrounding air chambers. Stomatal opening may also be produced experimentally by the removal of the CO_2 from the air passing over a leaf. This control by CO_2 helps to explain why stomata are normally open during the day and closed at night. The usual pattern of stomatal aperture (and of transpiration rate) is for a sudden rise to occur at dawn (Fig. 35), proceeding to a

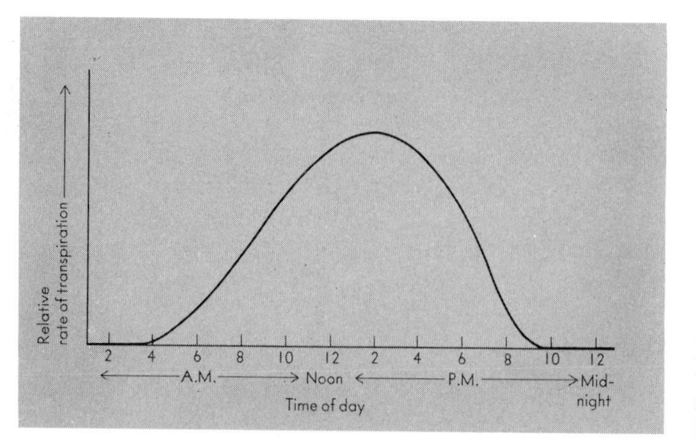

Fig. 35. The course of daily transpirational water loss. This curve tends to follow solar radiation and air temperature.

maximum near noon, followed by a temporary decline as the plant desiccates somewhat because of rapid transpiration, followed then by a small rise as the plant recovers from this desiccation, and, finally, by a decline as darkness approaches. Stomatal behavior and transpiration, therefore, are both governed by an interplay of water supply to the plant, by illumination, and by the CO_2 content of the atmosphere.

Plants vary widely in the number and distribution of their stomata. Plants adapted for life in arid regions (xerophytes) tend to have fewer stomata per unit area than do mesophytes; their stomata may also be located in sunken areas of the highly cutinized leaf or stem surface, thus minimizing water loss by restricting air turbulence. In some mesophytes, stomata are abundant on both surfaces of the leaf; in others, stomata may be restricted to the lower surface. In the cucumber, there are more than 400,000 stomata per square inch of leaf, while in some grasses, the number may be less than 50,000. In any event, in a wide variety of plants it has been estimated that when fully open, the stomata occupy 1–3 per cent of the total area of the leaf and thus constitute essentially no barrier at all to the free diffusion of water vapor out of the leaf. This fact accounts for the tremendous quantities of water lost by well-watered plants growing in bright light in warm temperatures. Even within a single plant, leaves vary widely in the number and distribution of their stomata; "shade" leaves generally have fewer stomata per unit area than do "sun" leaves.

Ultimately, all the water evaporated from leaves and other aerial parts of plants comes from the soil, from which it is absorbed by roots. The soil may be considered a sort of water reservoir, which is alternately filled and depleted. After a rain, a front of free water percolates downward through the soil, which is then said to be at *field capacity*. At field capacity, water may be very easily removed from the soil by plant roots, or by mechanical means such as centrifuging or squeezing. As the soil desiccates progressively, a point is reached where the difficulty of removing water becomes so great that the plant cannot supply its aerial parts with enough water to prevent wilting. The percentage of water in the soil at this point (called the *wilting percentage*) varies widely from soil to soil, being low in coarse sandy soils and relatively high in fine clay soils. Irrespective of the type of soil and type of plant, water cannot be removed from adsorption on soil particles if more than about 15 atmospheres pressure are required for the release. Thus, although clay soils hold more water than do sandy soils, they also retain more water in a tightly bound condition, unavailable to the plant. As far as we know, water is absorbed by roots through purely osmotic forces and is taken up mainly by the young, growing parts of the root system.

The mechanism by which water is transported to the tops of tall trees has long puzzled plant physiologists, and is still not completely understood. Atmospheric pressure will raise water only about thirty feet, and the tallest trees known rise to more than ten times this height. The forces involved must therefore be equivalent to about ten atmospheres of pressure. The theory currently most favored by plant physiologists is the so-called *transpiration-cohesion-tension theory.* This theory supposes that the evaporation of water from mesophyll cells sets up a diffusion pressure gradient for water that is communicated from cell to cell, until the dead xylem elements are reached (Fig. 36). The "pull for water," when trans-

Fig. 36. The "transpiration-cohesion-tension" theory of water rise.

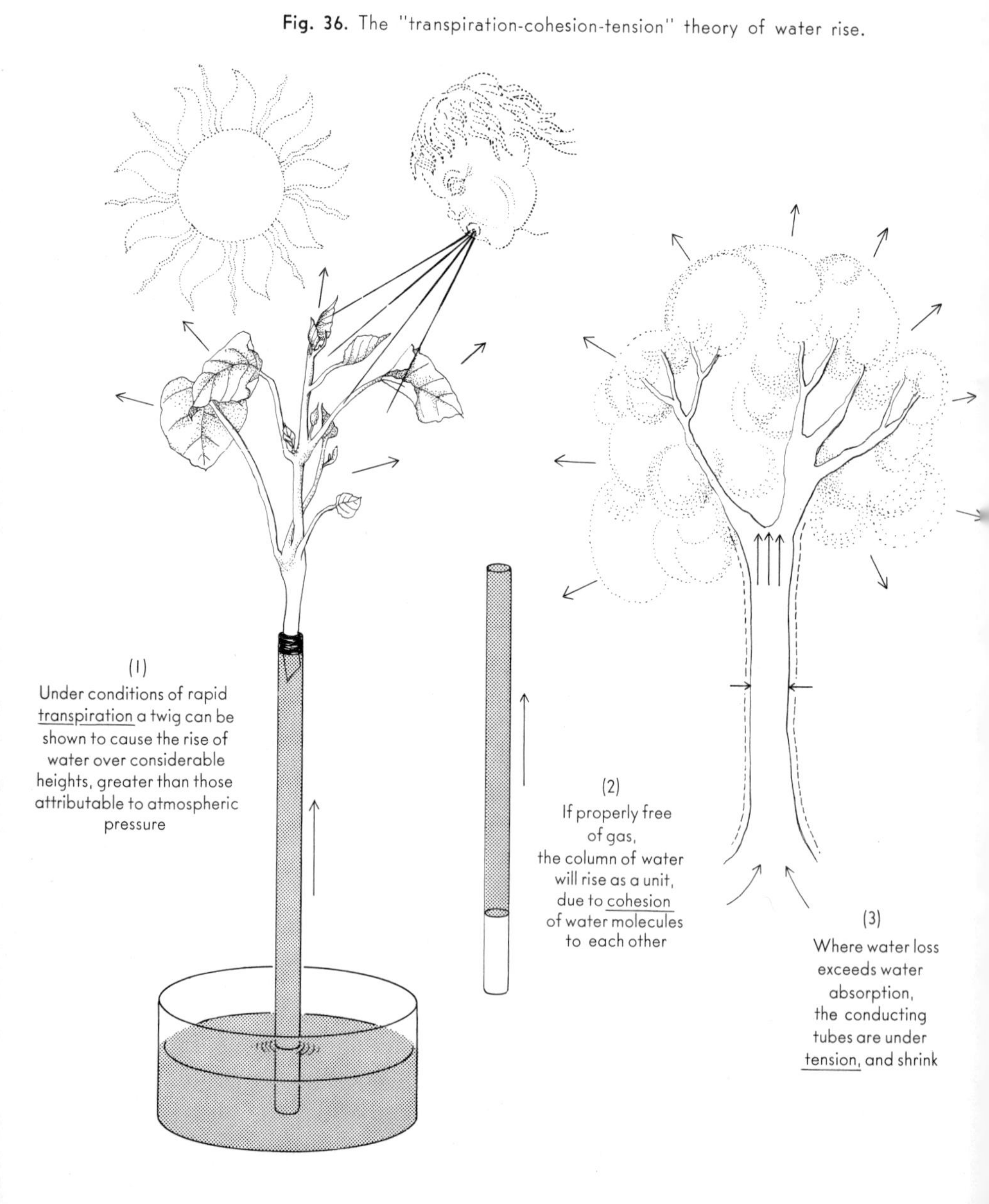

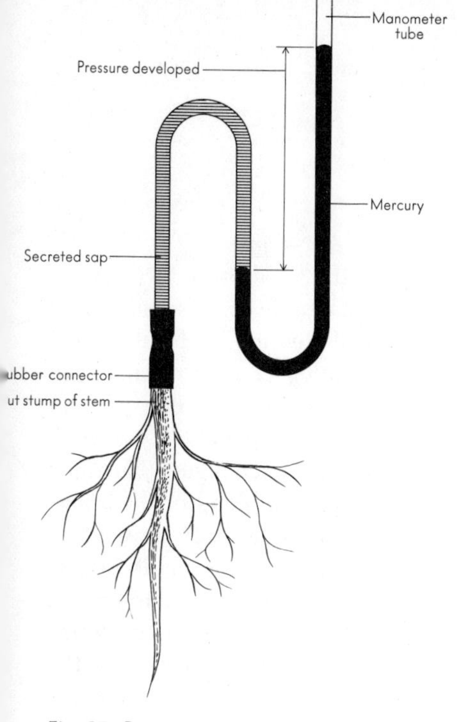

Fig. 37. Roots may secrete water against a large pressure gradient.

mitted to an open-pipe system such as tracheids and vessels, results in a mass lifting of a water column up these pipes, which are therefore placed under a condition of tension. Thus, according to this theory, the motive force for the rise of water in xylem is the evaporation of water molecules from the leaf, and water rise is independent of the vital activities of any cell. This theory is supported by at least three classes of evidence: (1) actively transpiring shoots can be observed to develop lifting pressures of several atmospheres; (2) completely inorganic models can be shown to work in the same way; and (3) the killing of stem cells by steam or poisons does not interfere with their capacity for water transport.

Another theory, the so-called root pressure theory, attributes greater importance to a pressure from below. Supporters of this view point out that when stems of many plants are severed just above the ground line, large quantities of fluid are exuded from the cut surface. If a manometer (a pressure-measuring device) is attached to such a cut stem, the roots can be seen to produce a pressure of several atmospheres (Fig. 37). Root pressures of appropriately high magnitude, however, have never been observed in the roots of tall trees, and, in any event, root pressures seem to be lowest when transpiration rates are highest. Most botanists are therefore willing to attribute to root pressure at best a subsidiary role in the process of water rise in trees.

4

Plant Growth

The development of the seed into the mature plant is a remarkable process, involving growth by cell division and cell extension, differentiation of new organs such as roots, stems, leaves, and flowers, and a long series of complicated, yet well-integrated, chemical changes. The final form of the plant is a blend of the inherent genetic patterns already present in the fertilized egg, plus the modifying effects of the forces of the environment. While the former set the ultimate limits between which the plant may vary, the latter determine where within these limits the developmental pattern shall be set.

The seed contains an embryo plant, surrounded and protected by a seed coat and supplied with a source of stored food (Fig. 38). The plant embryo is a bipolar axis, containing a young root growing point and a shoot growing point. The cotyledons, or embryonic leaves, are attached laterally near the midpoint of this bipolar axis. In some instances, the cotyledons are elongate, thin, and leaf-like, serving to digest the stored food of the endosperm tissue, and expanding into leaf-like photosynthetic organs after completion of this task. In other instances, they are fleshy storage organs, above or below ground, that have absorbed the endosperm prior to maturation of the seed; such cotyledons rarely become leaf-like or photosynthetic.

At the onset of germination, the seed absorbs large quantities of water, and the hydration of the cells at the growing point stimulates them into

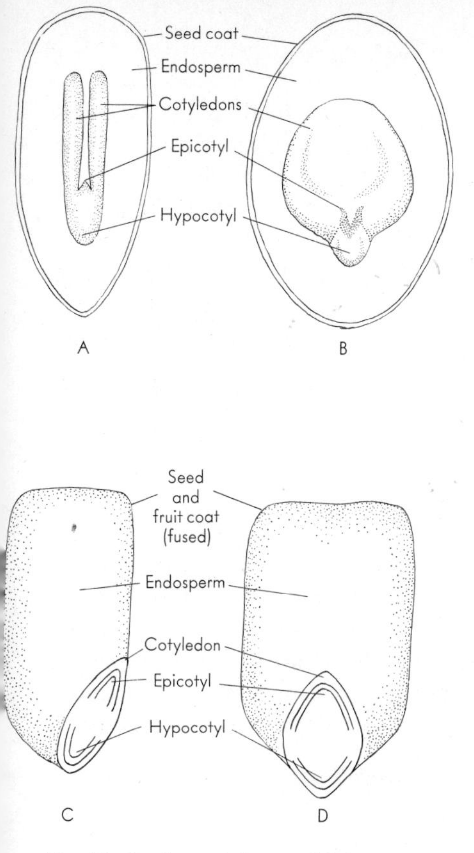

Fig. 38. Seeds contain a miniature plant (*embryo*), sufficient stored food to get the plant started (*endosperm*), and a protective layer (*seed coat*). (A) and (B) represent sections perpendicular and parallel to the flat surface of the dicotyledonous seed. (C) and (D) are the same for a monocotyledonous fruit (*grain*), containing a single seed.

mitotic activity. For reasons we do not understand, the root almost invariably begins development first and is followed later by growth activity at the shoot growing point. At both apices, growth is due to the formation of new cells by the *meristematic* (dividing) areas of the growing point, followed by elongation and differentiation of these cells. In the root, the processes of cell division, elongation, and differentiation occur in fairly well-defined regions that overlap considerably. Since the root must grow downward through a firm and resistant soil medium, its tender growing point requires protection against abrasion. This is furnished by a group of cells referred to as the *root cap*. This cap is produced by divisions of the meristem and is continuously flaking off and being replaced.

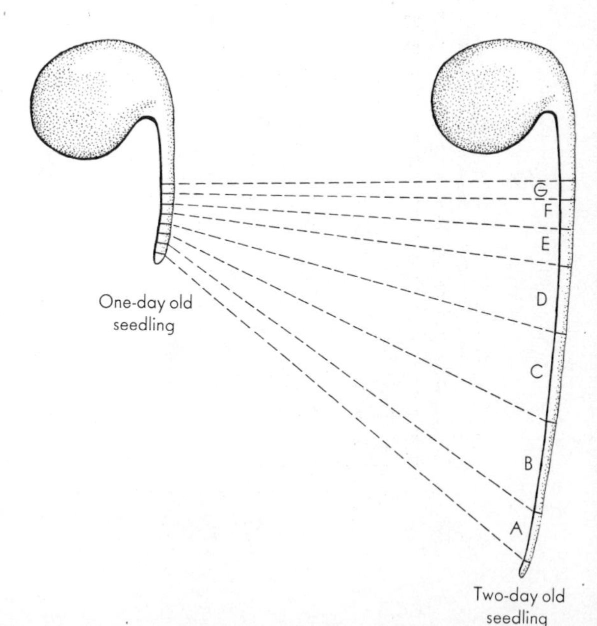

Fig. 39. Detection of zones of elongation by the parallel-line marking technique. Zones B, C, and D immediately behind the apex have elongated the most.

One marked difference between plants and animals is the fact that plants grow in restricted and localized areas near the meristems, while animals tend to have growth zones distributed all over the body. The restricted growth in plants can be visualized by the simple technique of marking the surface of the root or stem with equidistant lines of some non-toxic substance, such as charcoal in lanolin paste (Fig. 39). After some days, it will become apparent that the area just behind the tip is the region where the most rapid growth occurs. This is the region of cell elongation. Clearly, cell division itself does not do much to increase the size of the plant body; what it does is provide new units of potential elongation, which enlarge at some later time when they are farther back in the organ containing them.

The Kinetics of Growth

If the logarithm of the size of the plant is plotted as a function of time in days after germination, the curve of Fig. 40 results. This S-shaped, or sigmoid, curve is typical of the growth of all organs, plants, populations of plants or animals, and even of civilizations of men. It can be shown to consist of at least four distinct components: (a) an initial *lag period* during which internal changes occur that are preparatory to

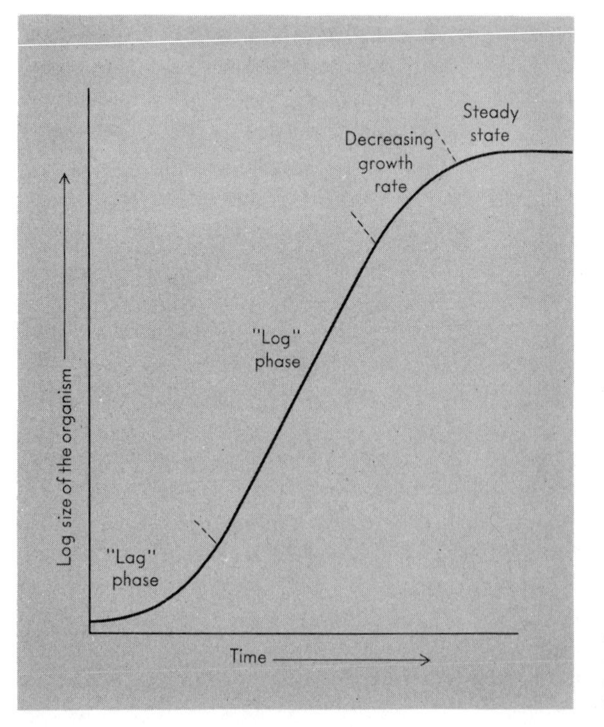

Fig. 40. The sigmoid growth curve. This curve is characteristic of single cells, tissues, organs, organisms, and populations.

growth; (b) a phase of ever increasing rate of growth; since the logarithm of growth rate, when plotted against time, gives a straight line during this period, this phase is frequently referred to as the *log period* of growth; (c) a phase in which growth rate gradually diminishes; and (d) a point at which the organism reaches maturity and growth ceases. If the curve is prolonged further, a time will arrive when senescence and death of the organisms set in, giving rise to one or two other components of the growth curve.

With plants, senescence and death are apparently not a necessary part of the developmental cycle. We know, for example, that some pine trees and Sequoia trees of the western United States attain ages of well over 3000 years. That they die eventually is due probably to the onset of infection or to the weakening of the mechanical base from which they spring. If such adverse conditions are somehow prevented, growth can continue potentially indefinitely. Indeed, the techniques of tissue culture, in which parts of plants are excised and placed upon artificial media, have been employed to demonstrate the potential immortality of plant cells. For example, in the year 1937 several research workers in France demonstrated that portions of carrot roots could be removed aseptically and placed on properly defined chemical media, resulting in the production of a rapidly growing undifferentiated mass of callus tissue. If such tissues are subdivided and transferred to new flasks at frequent intervals, their growth continues at a constant rate and shows no sign of diminishing even after 20 years. A carrot is normally a biennial plant, and the plant from which the original carrot tissue was taken certainly must have died many years previously. The normal cessation of growth in a plant, then, must be due to the onset of some inhibition which, if removed or circumvented, could result in a constantly growing plant that would be potentially immortal.

An analysis of growth curves has thus furnished us evidence for the existence of physiological growth controls of various kinds. The length of the lag period, for instance, can be used to analyze the nature of the changes which must occur prior to the inception of growth. With many seeds the lag period is only a few hours long; with others it may be days, weeks, or even months in length. In the latter case, inhibitory substances are probably present in the tissue, and growth cannot begin until these substances have been destroyed metabolically or removed by leaching or by other means. The rate of growth during the log phase is often determined by hormonal substances that we will discuss later in this chapter. An analysis of the slope of this curve can frequently tell us much about the genetic background of the plant's growth potentiality, as well as the adequacy of the environment in which the plant is growing. The total height of the plant and the time of the onset of the stationary phase is also frequently genetically controlled, but is definitely susceptible to control by the environment. Finally, the senescence and death of the organism are not necessary consequences of the genotype of the organism, but are under the control of the experimenter.

Meristems and Tissue Organization

In the highly organized context of a plant root or shoot, each cell goes through an orderly series of developmental phases. The cubical cell produced in the meristematic region of the growing plant is multivacuolate, and its increase in size, especially in length, is mainly a consequence of the uptake of water into the vacuoles. As the many separate vacuoles grow in size, they ultimately fuse into one large central vacuolar tank. The rest of the cell keeps pace with the increase in size by a synthesis of cell wall material, cytoplasmic material, and the various types of cell organelles. The major part of extension growth and increase in fresh weight is thus accomplished in the area of *cell elongation.* Concurrent with elongation, but sometimes subsequent to it, differentiation occurs. The cells on the exterior of the root, for example, adopt one of two final forms. They are either flattened epidermal cells or they are root hair cells. The latter consist of epidermal cells with tremendously elongated filamentous protuberances, which are very effective in the absorption of water and minerals. During the rapid growth of such cells, the nucleus is almost invariably found at the tip of the growing root hair and appears to be the center of great metabolic activity. Root hair cells are relatively short-lived, are present in great numbers, and are rapidly produced in large numbers as the root tip continues to push its way through the soil. These root hairs greatly increase the surface area of the external medium that is brought into intimate contact with the root.

The central tissues of the root differentiate into the vascular elements, since roots characteristically contain no pith (Fig. 41). The same pattern occurs even in tissue culture where it is rare to see tracheids or vessels differentiating on the surface of a mass of callus tissue. However, deep within the rapidly growing masses of an undifferentiated tissue culture may be found little whorls of nonfunctional but morphologically well-differentiated tracheid elements. We therefore suppose that there is something about the interior of a mass of tissue that favors the xylem type of differentiation. This "something" could be either the anaerobic conditions found deep within a mass of cells or, conversely, a lack of contact with the medium. Around root xylem cells are found bundles of phloem, the meristematic *pericycle* that gives rise to branch roots, and an *endodermis* that surrounds the entire central vascular cylinder. The endodermis has a curious structure, the *Casparian strip,* which represents a band-like thickening of the radial walls of the cells of the endodermis. Some investigators think the water-impervious Casparian strip represents a sort of water dam that prevents diffusion of water along the wall and forces

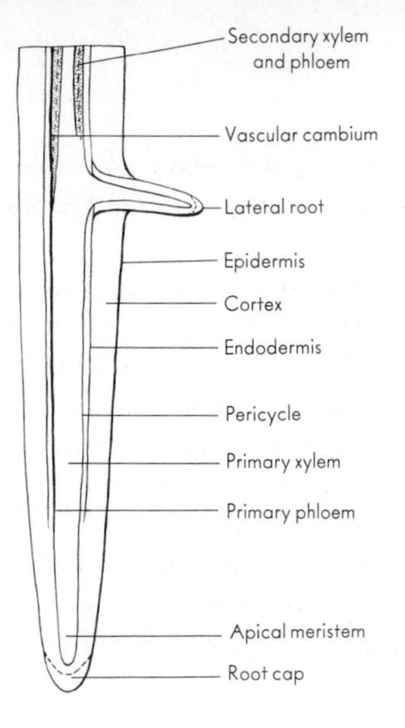

Fig. 41. A longitudinal section of a growing root, showing major zones. (From John G. Torrey, "Experimental Modification of Development in the Root," in *Cell, Organism and Milieu*, edited for The Society for the Study of Development and Growth by Dorothea Rudnick. Copyright 1959 The Ronald Press Company.)

movement of all materials through the differentially permeable membrane of the endodermal cells. This theory is still somewhat in doubt.

Between the internal vascular cylinder and the epidermis lies a group of loosely packed cells called the *cortex* (Fig. 42). These cells are parenchymatous in nature, are large and thin-walled, and contain obvious nuclei and large central vacuoles. Their function is presumably to store reserve materials in the root. As a tangentially dividing *cambium* develops between the xylem and the phloem of the central cylinder and as the roots thicken because of the activity of this cambium, the cortex becomes successively smaller and smaller because it cracks and is flaked off at the periphery. Finally, in an older root, the epidermis and cortex are

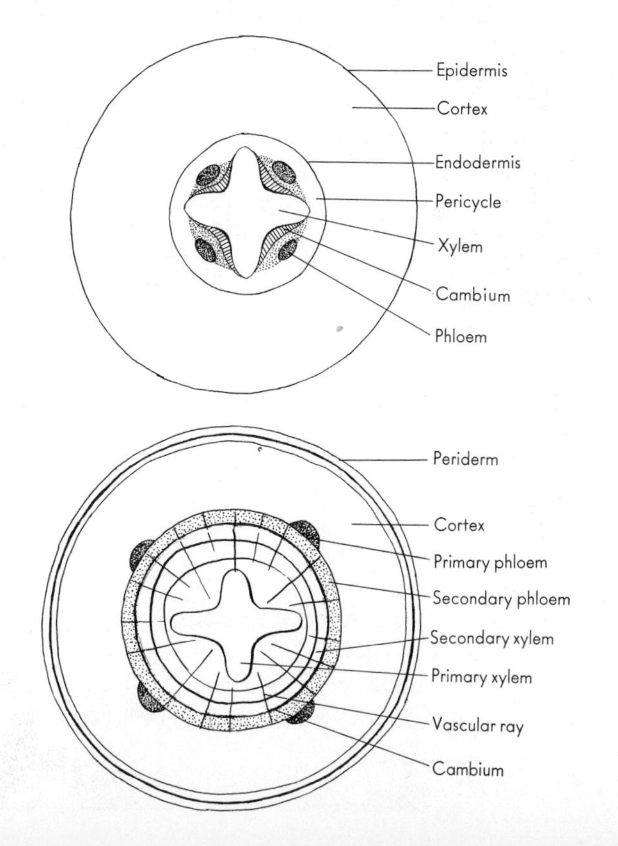

Fig. 42. (Top) Cross-section of a root without secondary thickening. (Bottom) The same root after considerable secondary growth.

completely lost; the external layer is derived from a secondary meristem, the *cork cambium*, which gives rise to corky cells at the exterior known as *periderm tissue*.

This pattern is even more clearly apparent in stems that thicken as they age. A stem apex, like the root apex, consists of a meristematic zone of cells that are in rapid division (Fig. 43). Posterior to the cell division area is a region of rapid cell elongation that overlaps with and merges into the region of cellular differentiation. Events at the stem apex are somewhat more complicated than at the root apex, because, in addition to stem tissue, leaves and buds are periodically formed by the apical growing point. The buds are first visible as minute projections of tissue and, depending on their structure, give rise ultimately to a vegetative bud or to a flower bud. The nature of the bud is, in many plants, controlled by en-

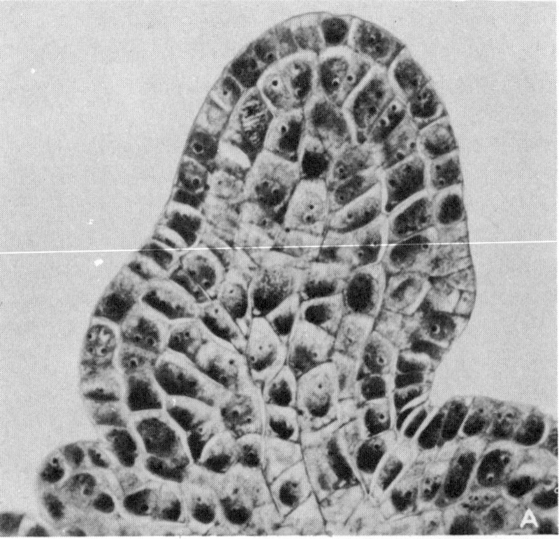

Fig. 43. (A) A median longitudinal section through the apical meristem of wheat. The swelling at the left is a floral primordium; the smaller one at the right will form a leaf-like organ, the lemma, which subtends the flower. The larger swellings below are older primordia. (B, C, D) Dissections of successively older wheat meristems which will produce inflorescences. Note the development of more and more surface ridges which eventually become separate, mature organs. (Courtesy Dr. C. Barnard and the Division of Plant Industry, Commonwealth Scientific and Industrial Research Organization, Australia.)

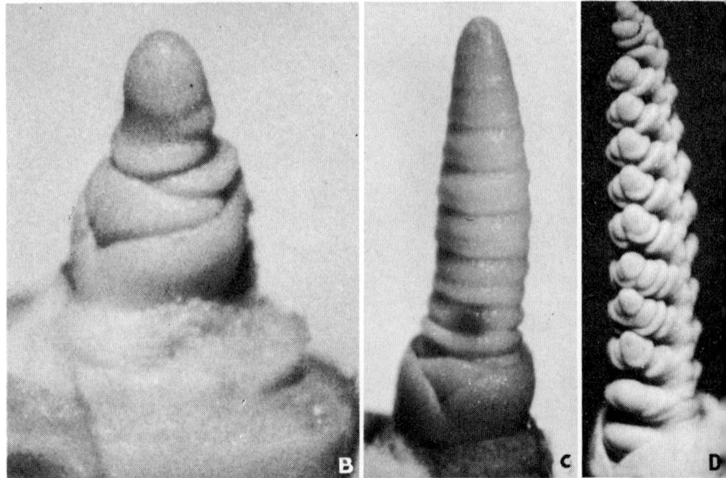

vironmental factors such as temperature and light conditions. These will be discussed in a later chapter.

Posterior to the region of elongation is the region of differentiation. Here, too, one can see quite clearly the development of epidermal tissues, of a central vascular cylinder, and of cortical cells between the two. Probably the main anatomical difference between stems and roots is the fact that stems generally possess a central pith; the xylem is found around the pith and the phloem around the xylem. Stems grown in the light generally do not have an endodermis; stems grown in the dark usually do. In stems, as in roots, the cambial layer develops between the xylem and phloem, and, by rapid divisions in both directions, it gives rise to cells that differentiate into xylem elements toward the interior of the cambium, and to cells that differentiate into phloem elements toward the exterior. The stem thus grows in girth because of an internal expansion. Ultimately, great pressures are set up by this growth in girth from within, and the external layers crack and flake off. As they do so, the plant produces new protective cells under the areas that flake off. Here, again, it is a *cork cambium* that appears, and the cells produced by this cambium are the heavy-walled waterproof cells typical of corky tissue. With continued cambial growth and an increase in girth, the successive layers of cork, which are now part of the bark of the tree or shrub, will flake off and be replaced in turn by new corky cells produced from existing or newly formed cork cambium layers (Fig. 44).

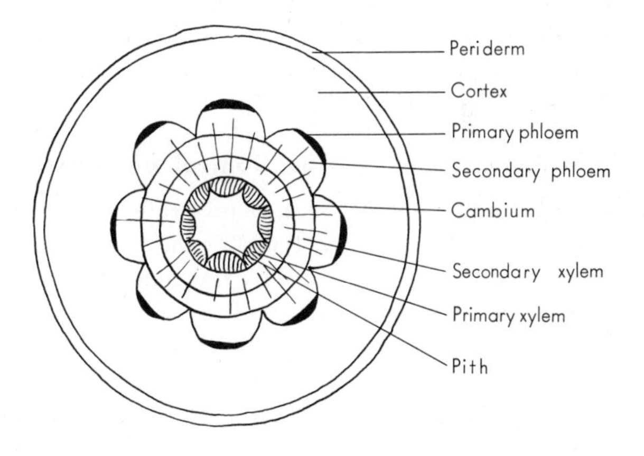

Fig. 44. A herbaceous stem with some secondary thickening.

The annual rings characteristic of the stems of trees result from the different climatic conditions in different periods of the year. In the spring, when the supply of water is ample and other conditions are favorable, the cambium produces cells that are relatively thin-walled and that contain a large central lumen. In the summer and fall, when the water supply and other environmental factors tend to be less optimal, the tracheids that are formed have thicker walls with smaller lumina. This regular alternation of spring and summer wood produces the annual ring (Fig. 45). Generally, the transition from spring to summer wood is gradual, while the abrupt halt at the end of the growing season is sharply delineated from the spring wood of the following year.

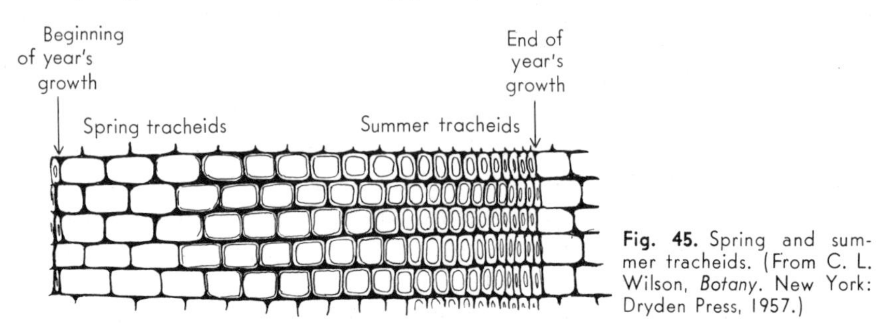

Fig. 45. Spring and summer tracheids. (From C. L. Wilson, *Botany*. New York: Dryden Press, 1957.)

The regularity of annual rings has enabled us to date trees and also civilizations in which remains of trees have been found. For instance, we know that certain climatic cycles have occurred in various regions. If a year is especially favorable to growth, a very thick annual ring is produced, while in drought years very small annual rings appear. Thus a piece of wood forming the supporting timber of a house in a now extinct civilization could be collated with other materials whose age is known and the civilization dated by this technique. Although extremely useful, the method is not infallible, because trees sometimes produce several growth rings in one year, and the annual rings of successive years are not always sharply separated.

The Control of Growth

As previously stated, a plant's rate of growth depends on both the genetic constitution (genotype) of the plant and on the various factors in its environment. The control by the genotype may be seen in experiments that compare the growth rate of several corn inbreds with that of the hybrid produced by cross-breeding them. Geneticists have known for many years that successive inbreeding diminishes the vigor of the stock, but cross-breeding between weakened inbred lines often produces a very vigorous hybrid. This condition is called *hybrid vigor* or *heterosis*. The causes of heterosis are not yet clearly understood, but the lack of vigor in inbreds is believed due to the accumulation of harmful recessive genes. Thus, if hybrids and inbreds are grown together in the same environmental complex, the hybrid will consistently outstrip the inbreds, indicating that their genetic capacity to utilize the forces of their environment is different from those of the inbred.

With any given genotype, tremendous control over growth may be exerted by obvious influences in the environment. For example, water deprivation will result in a greatly diminished growth rate. Similarly, a deficiency of nitrogen, potassium, phosphorus, or any of the other elements may result not only in the cessation of growth but in the death of the organism. The intensity of light falling on a photosynthesizing organism may determine not only its growth rate, but whether or not it will survive. The absorption of quanta by the photosynthesizing surface must be adequate to insure a rate of CO_2 fixation and energy storage commensurate with the needs of the developing organism.

Finally, the temperature of the environment may be extremely important. With most chemical processes, the rate of the reaction increases steadily with an increase in temperature. In fact, over most ranges, the rate of a chemical reaction is doubled by an increase in temperature of 10°C, or, in more technical language, the temperature coefficient (Q_{10}) is equal to 2 (Fig. 46). With physical reactions, such as diffusion, the rate is increased only slightly, perhaps on the order of 10 or 20 per cent by an increase in temperature of 10°C. The Q_{10} for such an effect is therefore 1.1 or 1.2. Plants grown in temperatures over the range of 0–30° show a steady increase in growth with increase in temperature, the Q_{10} approximating 2 or even more. For reasons that we do not understand, different plants have vastly different temperature optima, indicating that some fundamental biochemical process in their make-up has a great sensitivity to environmental temperature. With all plants, a temperature is eventually

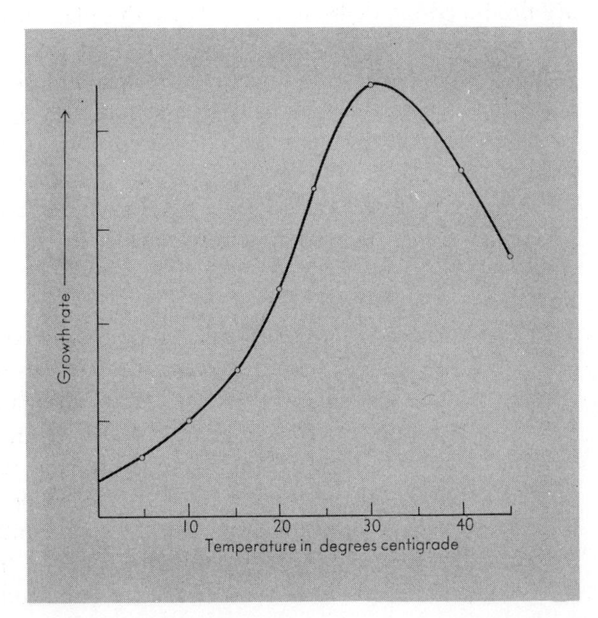

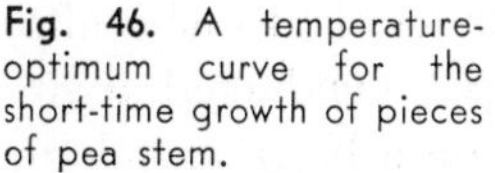
Fig. 46. A temperature-optimum curve for the short-time growth of pieces of pea stem.

obtained that is optimal for growth, and if the temperature is raised above this point the absolute growth rate declines, sometimes very dramatically. For most plants, this optimum is in the region of 28–32°C.

We do not know why plants are injured by temperatures of about 35°C, especially since the enzymes obtained from plants are, as far as we can tell, not damaged by this temperature. One guess is that certain essential chemicals which are required for growth, and which are produced by the growing plant cell, may either be destroyed very rapidly or not produced in adequate quantity at the elevated temperature. For example, the red bread mold, *Neurospora,* is known to contain certain genes that are "temperature sensitive." The gene responsible for the production of vitamin B_2 or riboflavin in one strain of *Neurospora* works quite well when the organism is grown at low temperatures, but does not function well when the organism is grown at higher temperatures. At 35°, therefore, the organism has an obligate requirement for riboflavin, while at 25° it is apparently autotrophic for this material. This same general situation probably holds for higher plants. If we knew the cause of the temperature-induced decline in growth rate, we could presumably greatly improve growth at high temperatures by applications of the material in question.

Growth Hormones

In addition to the water, light, carbon dioxide, and the various materials absorbed by the plant from its environment, other chemicals are required for plant growth. These substances, called *hormones,* are generally needed in only infinitesimally small quantities, and in most instances are produced in adequate amounts by the plant itself. By appropriate experimental techniques, we can deplete their supply in the plant, demonstrate their existence, and deduce much about their nature. A hormone is a substance produced in very small quantities in one part of the organism and transported to another part where it produces some special effect. Two major classes of growth-regulatory hormones have been shown to exist in most, if not all, higher plants. These are the *auxins* and the *gibberellins,* to which we now turn our attention.

THE AUXINS

The auxins are a group of substances produced by the growing apices of stems and roots. They migrate from the apex to the zone of elongation, where they are specifically required for the elongation process (Fig. 47). If the tip of a rapidly growing stem is removed, the growth in the region below the cut will slow down very quickly, and within several hours or days, depending on the type of plant, it will come to a complete halt. If the removed tip is replaced, growth of the stem continues almost normally, showing that some influence emanating from the tip is conducted across the wound to the growing cells. If the tip is placed on a block of gelatin

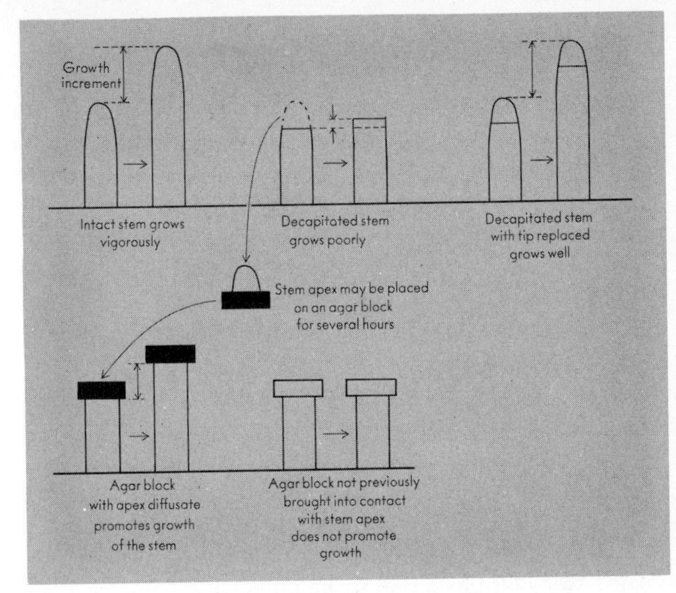

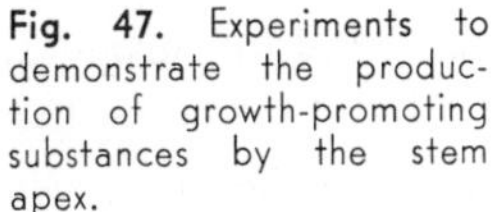
Fig. 47. Experiments to demonstrate the production of growth-promoting substances by the stem apex.

or agar for a period of several hours and the block without the tip is then transferred to the cut stump of a decapitated stem, the block will partially substitute for the tip in facilitating growth of the subjacent regions. From this experiment, we deduce that a substance, called *auxin,* moves from the tip to the block and from the block down to the base. Extensive chemical work of the last several decades has revealed the existence of many substances with auxin activity. Several of these have been isolated in pure form and have been shown to be native plant growth hormones. The most common of these substances is the simple material, indole-3-acetic acid (Fig. 48), which is probably derived from the amino acid, tryptophan.

The activity of indoleacetic acid may be shown very easily. If the growing portions of a stem, such as those of a pea plant, are excised and placed in petri dishes containing sucrose and a mineral salt solution, growth will be very slow. If, however, small quantities of indoleacetic acid are added, growth will be greatly enhanced, the effect being directly proportional, within certain limits, to the concentration of the auxin added. Usually an optimum concentration is attained, beyond which growth is once again somewhat slower, and ultimately growth may be inhibited completely (Fig. 49).

Usually the application of auxin to stems in an intact plant produces no extra increment of growth, from which we conclude that the stem is normally saturated with auxin produced by its own tip. In the root, on the contrary, the

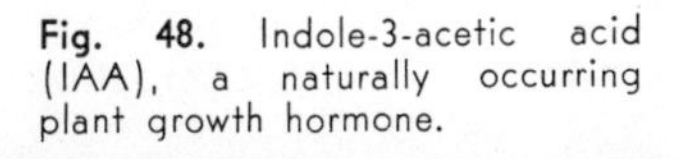
Fig. 48. Indole-3-acetic acid (IAA), a naturally occurring plant growth hormone.

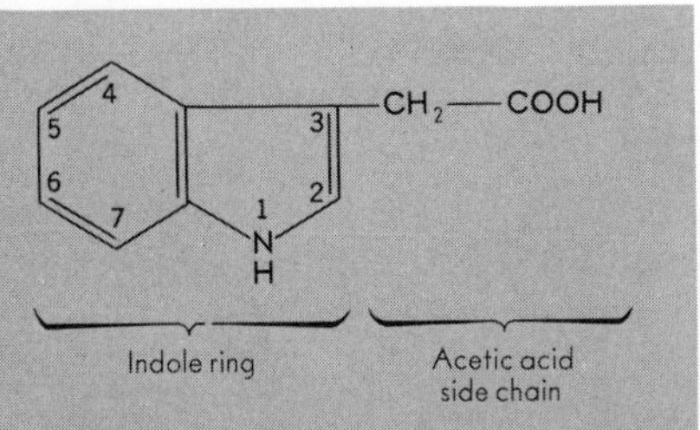

application of auxin usually results in growth inhibition, and in some instances the removal of the tip results in a growth enhancement. From this fact, we may deduce that the root normally operates under conditions of more than adequate auxin supply. Since the amounts of auxin in stem and root are about the same, the root must be more sensitive to auxin than is the stem. The growth of the root is promoted by concentrations of auxin lower than those which promote the growth of the stem and inhibited by concentrations lower than those which inhibit the growth of the stem.

The amount of auxin in any plant tissue may be determined by extraction with some solvent, followed by the application of the extract to some tissue that will respond quantitatively to the auxin contained in it. Normally, the tissue is placed in diethyl ether at a temperature near 0°C and gently shaken for a period of 2–4 hours. This ether extract is then concentrated and, when reduced to a small volume, is incorporated into an agar block that is then placed asymmetrically on the decapitated stump of the auxin-sensitive organ. Traditionally, the leaf sheath or *coleoptile* of dark-grown oat plants have been used. In this plant, the asymmetrically placed auxin enhances the growth of the tissue only directly below it. This unequal growth on the two sides of the coleoptile causes a curvature of

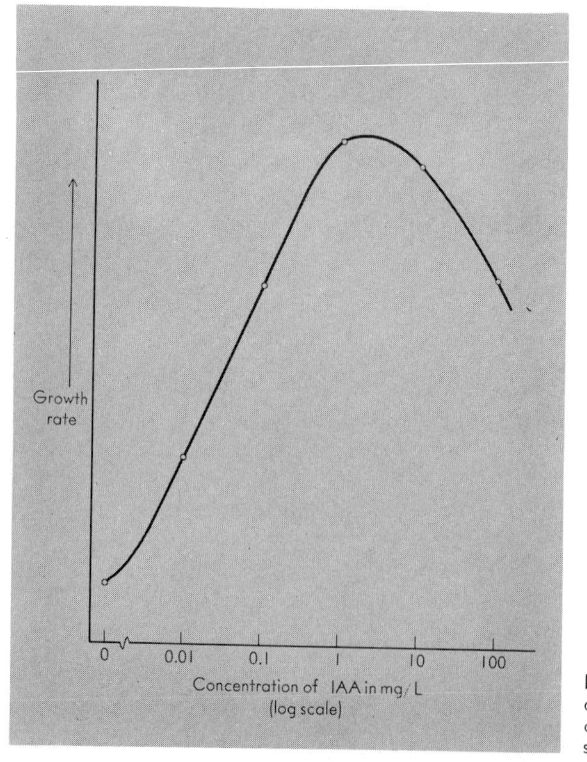

Fig. 49. A dose-response curve for the effect of IAA on the growth of pea stem sections.

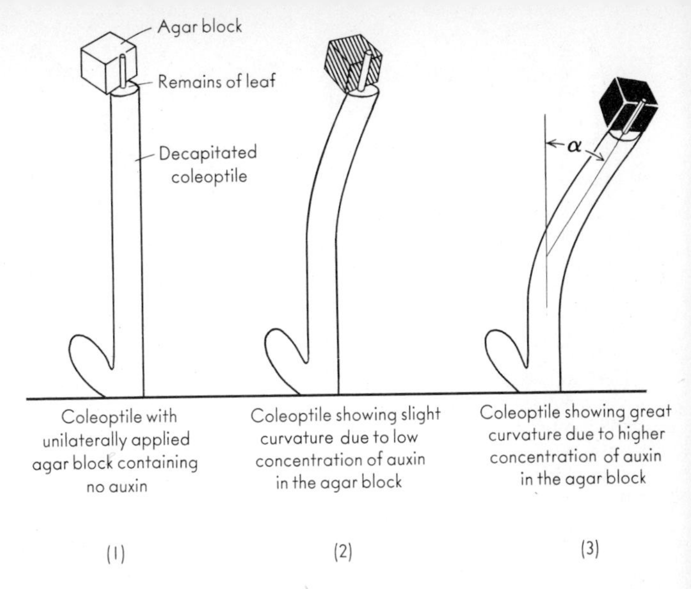

Fig. 50. The determination of auxin concentration by means of the oat coleoptile curvature test. The angle α is proportional to the auxin content of the agar block.

the organ, and is directly proportional to the amount of auxin incorporated into the block (Fig. 50). Thus, to determine quantitatively the amount of auxin in an unknown organ, the extract is made, the curvature is measured, and this curvature is compared with curvatures produced by known quantities of auxin in another experimental series. This technique of using the response of an organism to measure the amount of a chemical in an extract is called *bioassay*.

The curvatures produced by the unilateral application of auxin to plants bring to mind the curvature of various plant organs toward or away from light or gravity. In fact, such curvatures (called *tropisms*) are now known to be due to the asymmetrical distribution of auxin in the organ involved. For example, if an oat coleoptile is subjected to a low intensity of unilateral light, it will curve toward that light. This curvature results from the fact that the side near the light has had its growth somewhat depressed by the light, while the growth on the side away from the light has been accelerated (Fig. 51). If the tip of a unilaterally exposed coleoptile is removed and the amount of auxin of the two halves (light and dark) assessed by the curvature test described above, invariably the side away from the light will be found to have about twice as much auxin in it as the side toward the light. Plant physiologists have therefore concluded that light acts in producing curvature by affecting the distribution of auxin in the organ; this auxin concentration, then, controls growth.

Similarly, a stem laid prostrate can be shown, after some time, to accumulate more auxin on the lower surface than on the upper surface. This results in an accelerated growth below and an ultimate curvature upward. The growth downward of a prostrate root is a consequence of the different auxin sensitivity of the root. In the prostrate root, as in the prostrate stem, auxin accumulates below, but since in the normal root the

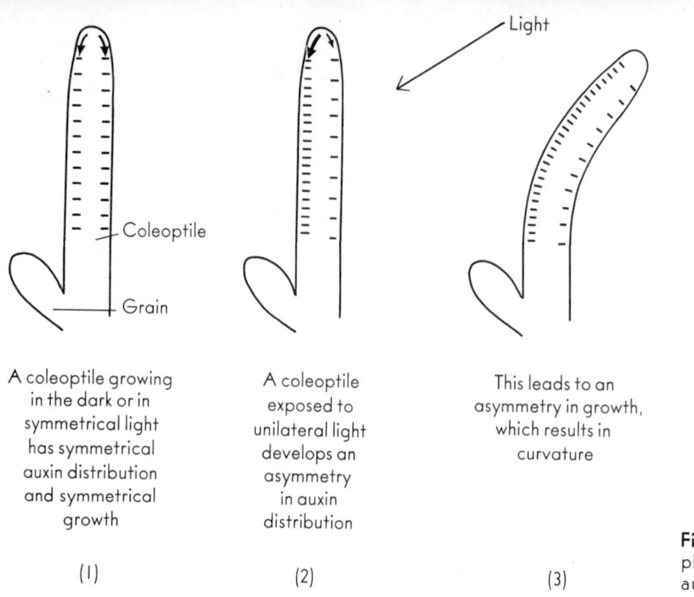

Fig. 51. An explanation of phototropism based on auxin distribution.

auxin concentration is already optimal, or supraoptimal, this greater concentration of auxin on the lower side leads to depressed growth, and thus to a downward curvature of roots.

Thus we see that auxin not only controls the over-all rate of growth, but also the direction of growth of the plant. We still do not know exactly how the asymmetrical distribution of auxin is brought about under unilateral light or gravitational stimuli. Some people believe that light and gravity actually cause a lateral migration of auxin formed in the tip, perhaps under the influence of electrical potentials set up by the stimulus. If this were true, it should be possible to demonstrate it by the application of radioactively labeled auxins. Experiments performed in this way indicate there is no influence of light or gravity on the transverse migration of such substances. An alternative and perhaps more attractive theory says that the stimulus affects the synthesis of auxin at the tip of the organ, perhaps by controlling the activity of the enzyme that transforms an auxin precursor to auxin itself.

The auxin produced in the apex migrates away from the tip at a rate approximating 1 cm per hour at normal temperatures. For reasons we do not completely understand, this migration is unidirectional, going only from apex to base (Fig. 52). The application of a block of agar containing auxin to the morphological apex of a stem, therefore, results in the rapid passage of auxin into the tissue, from which it may be collected by extraction or by diffusion into a basal block. If, however, a similar block is applied to the base of the same piece of tissue, it will not move through the base into the apex and out the end into a receptor block. There seems to be no morphological basis for this difference of action; rather there is some sort of physiological valve action involved about which we know

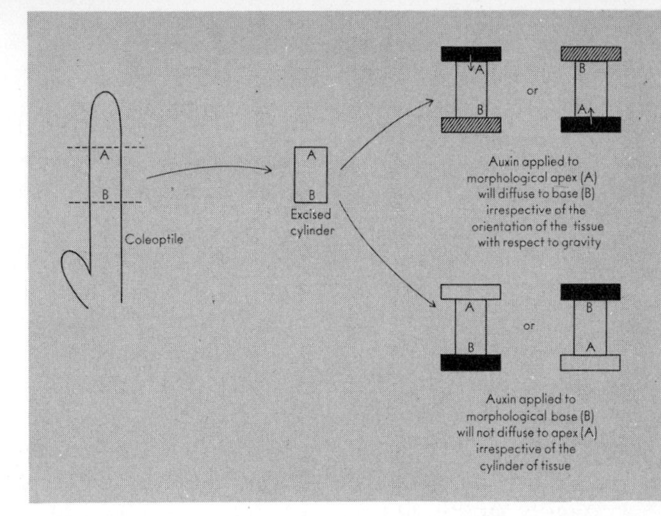

Fig. 52. The flow of auxin through coleoptile tissue is unidirectional.

very little. We do know that the polarity of transport may be interrupted by anesthetics such as ether and chloroform, but exactly what they are blocking is a matter of conjecture.

So far we have been discussing only the cell elongational aspects of auxin action. Auxin may also initiate or promote cell division activity. For example, in tissue cultures of normal cells excised from stems or roots, cell division is entirely dependent on the application of auxin to the medium. Similarly, the initiation of cambial activity in the spring in trees can be shown to be controlled by auxin diffusing downward from the developing bud. In addition, the formation of branch roots and adventitious roots from the pericycle region of stems or roots can be initiated by the application of auxins. In this mitosis-inducing activity, auxin apparently works together with other substances such as the constituents of nucleic acids. In fact, a substance called *kinetin* has been isolated from heated yeast nucleic acid and has been shown to be active in promoting the cell division activity of plant cells in the presence of auxin. There is, however, as yet no good evidence for the existence of kinetin or anything chemically related to it in plant tissues, but substances resembling kinetin in physiological action have been shown to occur in apple fruitlets and in coconut milk and other liquid endosperms.

In addition to its roles in promoting cell division and cell elongation, auxin has other correlative effects on the growing plant. For example, auxin is known to control the growth habit in the sense that it determines whether the apical bud or lateral bud on a stem will develop. In the intact stem of many plants, only the apical bud can grow. Removal of the apical bud, however, results promptly in the growth of one or several of the buds lower down. If the tip is removed and the cut surface covered with an auxin paste, the buds lower down will continue to be inhibited. From this type of experiment, it has been deduced that auxins inhibit lateral bud growth.

The mechanism of this effect is unclear, but the effect has nonetheless been put to some constructive use in agriculture. The potato tuber is a modified stem, and the "eyes" of the potato are actually buds, each subtended by a bud scale. The freshly dug potato is dormant, but after some months of storage the buds begin to sprout. The application to such potatoes of synthetic auxins results in a complete inhibition of this sprouting for very extended periods. This device has permitted the storage of potatoes for up to three years instead of one year. The synthetic auxin molecules (Fig. 53) that help produce this effect, and other agricultural effects to be listed below, are indolebutyric acid, in which the IAA side chain is lengthened by two carbon atoms, naphthaleneacetic acid, in which the indole ring is substituted by the naphthalene ring, and variously substituted phenoxyacetic acids, in which the phenoxy radical is substituted for the indole ring. These and many other substances are more stable than indoleacetic acid and are readily synthesizable in large quantities. They now furnish the basis for a large and growing industry based on the chemical regulation of agriculturally important plants.

Auxins are also important in regulating the fall of leaves and fruits from plants. The leaf blade is attached to the stem by means of a petiole that persists during the growing season but falls off at some time later in the year. As long as the leaf blade produces adequate quantities of auxin, the attachment of the petiole to the stem is firm. If, however, the leaf blade

Fig. 53. Various synthetic auxin molecules.

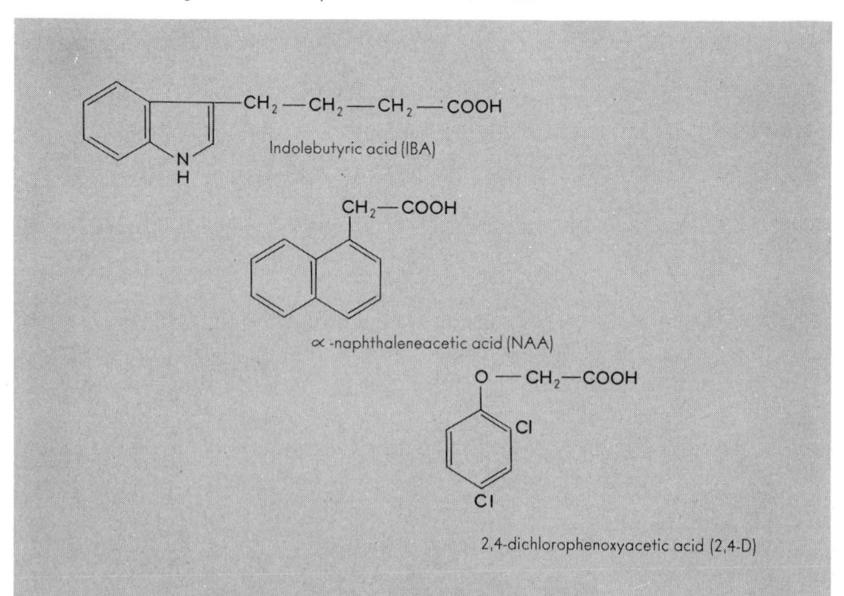

becomes deficient in auxin production, the petiole forms at its base a special layer of cells called the *abscission layer,* which is mechanically so weak that the leaf blade is easily caused to fall from the plant by a breeze or by mechanical irritation.

This knowledge has been put to good use in agriculture. For example, where it is desirable to retain leaves or fruits on a tree (as in apples and oranges), it is merely necessary to spray the tree with a dilute solution of 2,4-dichlorophenoxyacetic acid or some related auxin. This simple process has saved million of dollars for orchardists whose fruits normally fall off the tree when they are not yet ready for harvest. Similarly, by the production of chemical analogs that antagonize the action of auxin, leaves or fruits can be caused to fall prematurely from a plant. This fact greatly aids mechanical cotton picking. The mechanical cotton picker, which makes cotton production more economical and less dependent on human labor, does a good job as long as the plants do not have too many leaves on them. In order to insure that the condition of the plants will be correct, "anti-auxins" are sprayed over the cotton field several days before harvest. The machine then can run through the field very nicely, harvesting the bolls without too many interfering leaves.

In the pineapple plant, one auxin, α-naphthaleneacetic acid, has the remarkable effect of promptly inducing the onset of flowering. Although the mechanism of this effect is obscure, its value in pineapple agriculture is obvious. The plants can all be grown to a uniform size and naphthaleneacetic acid applied at any desired time. The fruits will then develop uniformly, making mass methods of harvest possible. The flowering of the litchi nut and several other fruits can also be controlled by the use of auxin sprays, but, unfortunately, this simple technique is not generally applicable.

In another important economic activity, auxin is applied to the pistil of the flower to produce artificial or *parthenocarpic fruits.* Normally, most fruits are formed as a result of pollination and fertilization of the ovary of the flower, and of the subsequent growth stimulation produced in the ovary by the fertilization process (Fig. 54). If, however, no pollen reaches the pistil, the development of the ovary into a fruit can be stimulated by the application of fairly large quantities of auxin-type materials. Naphthaleneacetic acid, for example, can be sprayed or daubed onto a tomato ovary to produce a fairly typical fruit that is large, red, and tasty, but lacking in viable seeds. So far, although it has been found possible to substitute auxin for the growth stimulus of pollination, it has not been possible to effect development of the egg without normal fertilization. The fact that auxin substitutes for pollination in stimulating development of the ovary wall suggests that pollination somehow results in a natural increase in auxin concentration in the ovary wall. This conclusion is borne out by auxin assay. Since the pollen itself does not contain so much auxin,

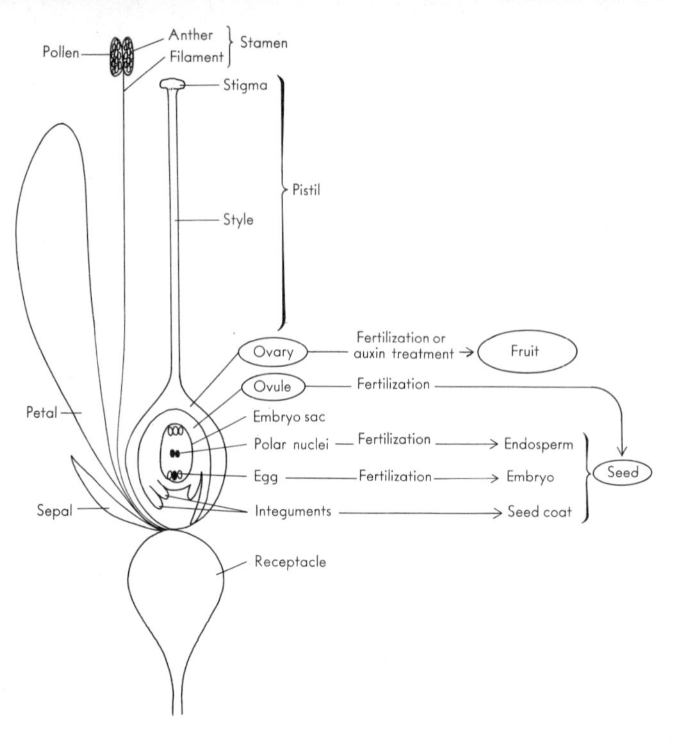

Fig. 54. The parts of a flower. The *ovary* develops into a fruit when stimulated by fertilization or an auxin spray. Development of an *ovule* into a seed occurs only after fertilization. In this process, the *egg* and *polar nuclei* are both fertilized by sperms produced in the pollen tube to form, ultimately, embryo and endosperm of the seed.

its function must be to activate the formation of auxin, probably by carrying into the ovary some stimulator of the enzyme that makes auxin from its precursors.

One final aspect of auxin action should be noted here. We have already described how supraoptimal concentrations of auxin produce marked inhibitory effects on the growth of certain kinds of cells. It is also true that some of the synthetic materials, such as 2,4-dichlorophenoxyacetic acid (2,4-D), have differential toxicity toward different plants. Thus the application of a given concentration of 2,4-D to a mixed population of plants may kill one type but not another. Fortunately, the spectrum of toxicity is such that this result can be exploited very effectively in agriculture. The 2,4-D shows very great toxic effect on dicotyledonous or broad-leaved plants, but is relatively non-toxic to monocotyledonous or narrow-leaved plants. In a lawn containing dandelions, therefore, the application of 2,4-D kills the dandelions while leaving the grass intact. Similarly, the application of 2,4-D to a cornfield containing the noxious bindweed will kill the bindweed, leaving the corn intact. As a result, chemical weeding has been substituted for the laborious, injurious, and expensive practice of mechanical cultivation. The savings from such practices are incalculable.

Before concluding our discussion of auxins, we should remark that the study of these compounds arose accidentally in the 1870's with the experiments of Charles Darwin and his son, Francis, who were investigating the curvature of grass seedlings toward light. By the use of small opaque capillary glass shields placed over the coleoptiles, Darwin was able to demonstrate that while only the tip of the coleoptile is able to perceive phototropically active light, a region some distance below the tip does the curving. He hypothesized in his book, *The Power of Movement in Plants,* published in 1881, that some stimulus passed from the tip down to the growing zone and exerted some specific growth effect. This idea spurred various investigators to further experiments that led in 1928 to F. W. Went's simple demonstration, by the technique mentioned above, of the existence of the diffusible growth hormone produced in the tips of oat coleoptiles.

This bit of history ought to teach us that in the search for knowledge the most impractical of investigations is frequently the most direct way to economically important results. Had Darwin been told by some supervisor to develop a weed killer, he undoubtedly would not have performed his perceptive experiments leading to the discovery of auxin. In the allocation of research funds, therefore, all opportunity should be given for the unfettered mind to operate as it most desires.

THE GIBBERELLINS

Another group of important plant growth hormones, also discovered through a series of accidents, is the gibberellins. In the last decade of the nineteenth century, Japanese rice farmers noticed the appearance of extraordinarily elongated seedlings in their paddies. They watched these seedlings closely, for any large-sized plant is usually considered by the alert farmer as a possible source of breeding stock for the improvement of general vigor. These tall seedlings, however, never lived to maturity and only rarely did they flower. The disease was aptly named Bakanae or "foolish seedling" disease. In 1926 a Formosan botanist discovered that these seedlings were all infected with a fungus called *Gibberella fujikuroi,* a member of the Ascomycete or sac fungus group (Fig. 55). If the fungus was transferred from an infected seedling to a healthy plant, the recipient became diseased. It was also observed that if the fungus were grown on an artificial medium in a flask, the nutrient medium accumulated some substance which when transferred to a receptor plant produced the overgrowth symptoms typical of the "foolish seedling" disease. This substance was named *gibberellin,* after the fungus which produces it.

In the 1930's Japanese physiologists and chemists worked together and succeeded in isolating from the *Gibberella* growth medium several substances, some inhibitory and some growth-promoting. The structural formula they finally proposed for the growth promoter, gibberellin, was somewhat in error, but nonetheless they had correctly diagnosed the general nature of the material and had produced crystals which when applied to test plants produced the typical hyperelongation symptoms of the

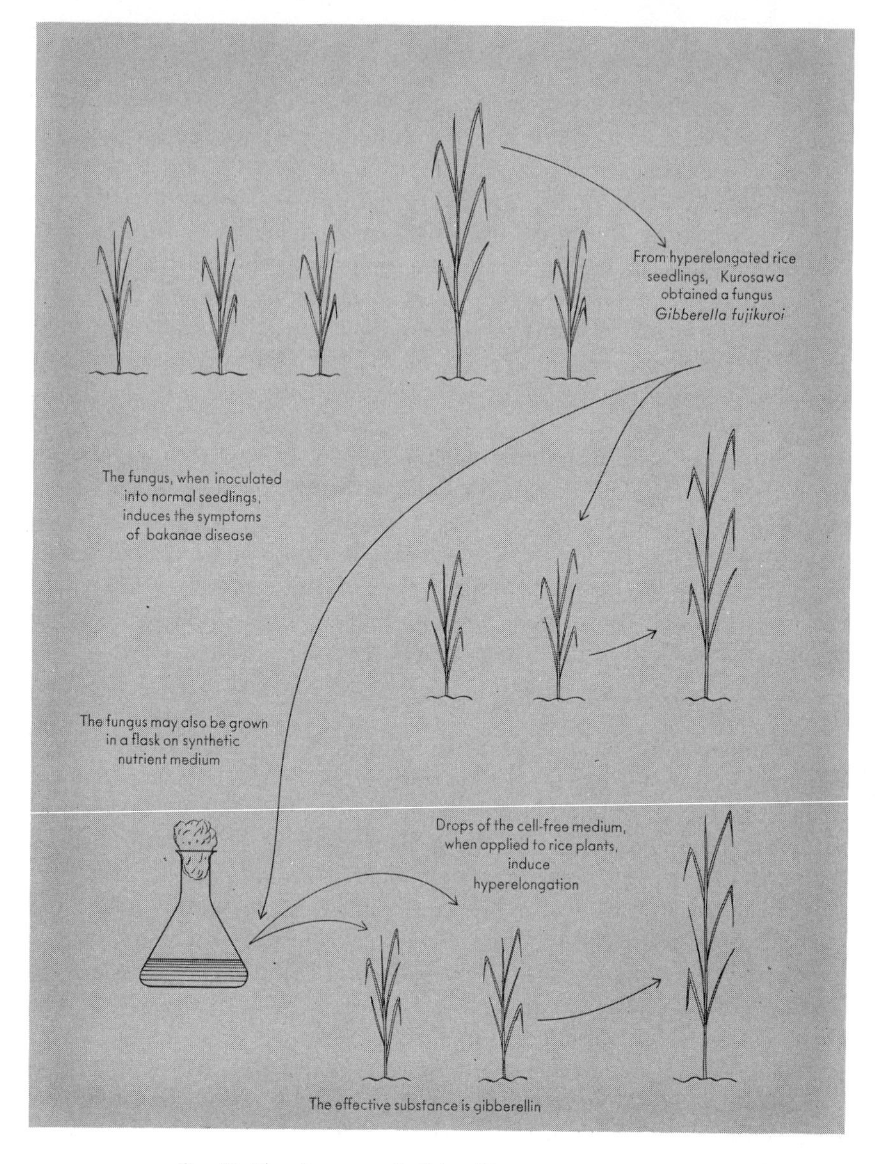

Fig. 55. The discovery of gibberellin.

Bakanae disease. This information was published in many papers before 1939, but, unfortunately, World War II intervened at this time and diverted the attention of most scientists to military matters. The existing story of gibberellin remained unknown in the Western world until about 1950, when several groups of workers, both in the United States and in England, uncovered these old papers and attacked the problem again.

By 1955 the Imperial Chemical Industries Laboratories in Britain had confirmed the original Japanese observation and had, in addition,

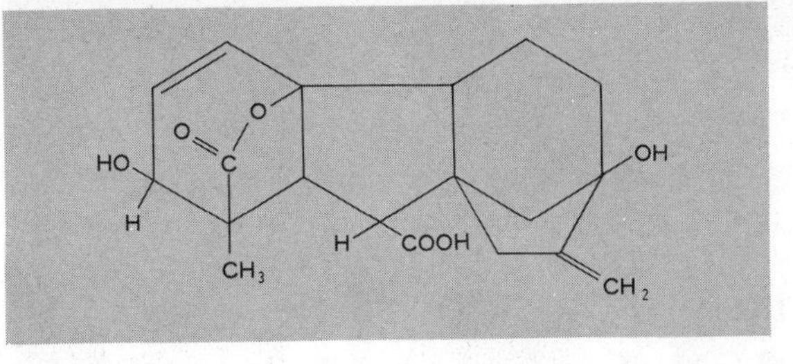

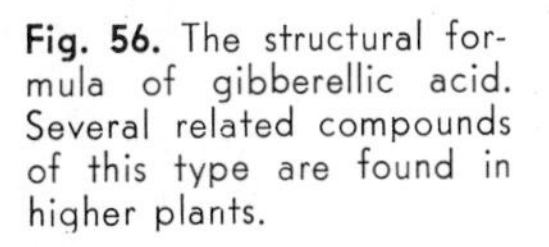

Fig. 56. The structural formula of gibberellic acid. Several related compounds of this type are found in higher plants.

isolated a substance they named gibberellic acid, which differed slightly from the material isolated by the Japanese (Fig. 56). Gibberellic acid, when applied to plants, produces tremendous hyperelongation effects on stems; in some instances, it also results in a diminution of leaf area. Its most dramatic effect is probably its prompt stimulation of flowering in a class of plants called long-day plants (Fig. 57). This phenomenon will be discussed in detail in the next chapter. It should be obvious that any chemical that causes a vegetative plant to initiate flower primordia is a potentially important one in agriculture. At the moment, the most important agricultural use of the gibberellins is in the grape industry, where the application of gibberellins to seedless grape clusters results in the retention and development of a greater number of grapes, and in much larger fruit (Fig. 58). For a relatively small expenditure of time and money on the chemical, the farmer obtains a much larger crop. Gibberellin is also

Fig. 57. The effect of gibberellic acid on the bolting and flowering of spinach (left) and cabbage (right). Each photograph includes a control on the left and a treated plant on the right. (Courtesy S. H. Wittwer and M. J. Bukovac, Michigan State University, and *The Hormolog.*)

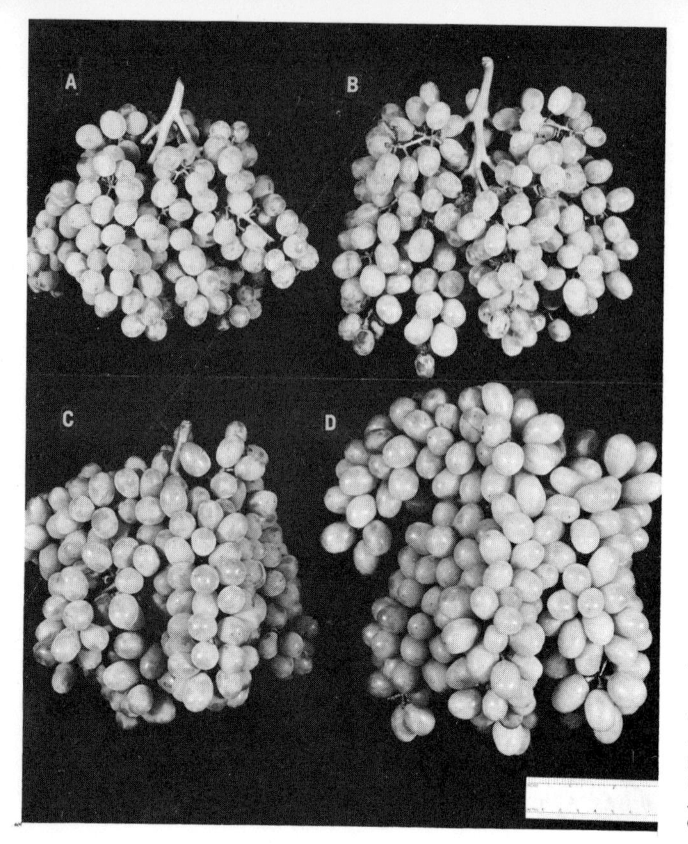

Fig. 58. The effect of gibberellin on the growth of Thompson seedless grapes. (A) Control grapes. (B, C, D) Grapes sprayed with 5, 20, and 50 parts per million, respectively, of gibberellin. (Courtesy R. J. Weaver and S. B. McCune.)

useful in celery growing, where it produces larger, more succulent plants in a shorter time. It can also be used to stimulate seed germination and growth of early seedling grass, including the barley that is used as malt in the brewing industry. Its potential in agriculture is very great, and is only now beginning to be explored.

Not all plants respond to gibberellin. In surveying the results of numerous tests, plant physiologists have noted that there is a correlation between the original size and vigor of a plant and the degree to which it responds. For example, if we compare dwarf peas with tall peas, or dwarf corn with tall corn, we note that the application of gibberellin to the dwarf causes the dwarf to assume the tall appearance, but application of gibberellin to the tall form results in little or no effect (Fig. 59). Since, in many instances, the difference between dwarfness and tallness resides in a single gene, the attractive hypothesis is suggested that dwarfness is actually due to the inability of the plant to produce enough gibberellin to satisfy its major needs. The application of gibberellin to certain genetic dwarfs would thus produce the tall appearance. Such dwarf plants treated with gibberellin would still, of course, have the dwarf genotype and when bred would yield dwarf progeny, even though they look tall.

If the hypothesis that dwarfness is due to lack of gibberellin is true, then tall plants should be demonstrably richer in gibberellin than dwarf plants. To assay for gibberellin in a plant extract, dwarf plants can be treated with known quantities of pure gibberellin to produce a "standard curve" of response. The extracts of tall and dwarf plants containing unknown quantities of gibberellin can then be introduced onto other test plants, and the amount of growth produced can be used, by reference to the standard curve, to estimate the amount of gibberellin present in the extract. If such assays are made on dwarf and normal plants, especially on the growing regions, no differences in quantity seem to appear. These experiments, however, have revealed that gibberellin-like materials do exist in plants, and this fact has encouraged chemists to attempt to isolate gibberellin from higher plants. So far, gibberellins have been isolated from the young seeds of various types of beans, where they appear to be present in quite large quantities, but all attempts to isolate them from growing vegetative regions of plants have been unsuccessful.

It is interesting that gibberellin would probably not have been suspected as a growth-regulating substance in higher plants had not an invading fungus, *Gibberella fujikuroi,* happened to produce enough of it to cause growth abnormalities and eventually the death of the infected plants. Once again, chance led to an important discovery.

Fig. 59. (A) Dwarf pea treated with water (left), and gibberellic acid (right). (Courtesy Anton Lang.) (B) Effect of gibberellic acid on normal and dwarf corn. Left to right, normal control, normal plus gibberellin, dwarf control, dwarf plus gibberellin. (From Bernard O. Phinney and Charles A. West, "Gibberellins and the Growth of Flowering Plants," in *Developing Cell Systems and Their Control,* edited for The Eighteenth Symposium of The Society for the Study of Development and Growth by Dorothea Rudnick. Copyright 1960 The Ronald Press Company.)

OTHER GROWTH PROMOTERS

In addition to the auxins and gibberellins, other promoters of the growth of plant cells are now known. For example, various workers have shown that coconut milk, the liquid endosperm of the coconut fruit, will greatly enhance the growth of numerous plants cells that otherwise grow with great difficulty. Extensive analyses of coconut milk have been made in an attempt to isolate the active substance, so far without success. It appears that coconut milk contains a mixture or variety of materials that together produce the growth-enhancing effect.

Another growth-promoting material is the synthetic substance, *kinetin,* discussed earlier in connection with the mitosis-inducing action of auxin. As far as we know, kinetin is not found normally in any plant. It is found in old, somewhat decomposed preparations of deoxyribose nucleic acid (DNA) or in fresh nucleic acid preparations that have been subjected to autoclaving or boiling. When kinetin is applied to certain plant cells, such as some calluses or roots grown in tissue culture, it acts together with auxin to produce a tremendous increase in mitotic activity. Auxin alone seems to produce only a swelling and an enlargement of the cell, while kinetin seems to favor mitotic activity. When auxin and kinetin are present in a reasonable ratio, the cells will grow and divide normally. It may be true that the normal mitosis-producing activity of auxin, alluded to earlier in this chapter, is a consequence of the interaction of auxin with some naturally occurring kinetin-like material. Such a material has been reported in apple and plum fruitlets, in coconut milk, and in tumorous plant tissues.

Still other growth-stimulatory materials have been described from time to time, but their over-all significance in plant physiology remains obscure. Usually green plants do not need external applications of vitamins, since they are autotrophic for these substances. There are in the literature, however, a few examples in which plants grown at abnormally high or at abnormally low temperatures respond with increased growth to the application of thiamin or other members of the vitamin B complex. Some workers have also found that adenine (a nucleic acid constituent) significantly enhances the growth of leaves, both on and off the plant.

Growth Inhibitors

In addition to substances that accelerate growth, plant tissues may contain relatively large quantities of substances that inhibit or prevent growth. In many instances, these substances appear to exert normal regulatory control over the growth patterns of the plant. For example, many seeds will not germinate if placed in the usual fashion on moist filter paper in a petri dish at room temperature. However, if the seeds are placed in running tap water for a period of several hours, they are stimulated to germinate promptly. It can be demonstrated that this effect is due to the leaching, out of the seed coat or other peripheral layers, of substances that normally prevent the germination of the seed. This process can be an important survival mechanism for seeds of plants living in arid zones. For example, on the desert, if a light rain were to fall, such seeds would not germinate. This is fortunate, because germination would produce a tender seedling that would die in the absence of additional moisture. When, however, a really heavy rainfall comes, the inhibitor is leached out of the seed, germination promptly occurs, and the plant can survive because of the abundance of water in the soil around the root of the tender germinating seedling.

In a similar fashion, inhibitors can be affected in seeds and buds by appropriate temperature treatment. It now appears that inhibitory substances are made during the growing season and accumulate in the bud, causing it to become dormant. The dormancy is normally broken by the destruction of the inhibitor when the plant is exposed to a sufficiently long cold period. The survival value of this mechanism is obvious. If, for example, a few warm days come very early in the year, an unprotected bud might start to grow and then be killed by a recurrence of cold weather. If, however, the bud is kept in check by the inhibitor, it will not grow until later in the season, when a sufficient number of cold days would have insured the complete destruction of the inhibitor. Such buds and seeds, then, contain a "chemical clock," and they tell time and record the passage of the winter months by means of a mechanism for the progressive destruction of the inhibitory substance.

Some of these inhibitory substances have been isolated and characterized. They belong to a wide variety of chemical types, from fairly simple phenolic substances to complex molecules. The plant seems to be rather versatile in fabricating the components of its chemical clocks.

We have already referred to the fact that artificial inhibitors of plant growth can be synthesized in the laboratory and that these inhibitors are useful in the control of weeds or other noxious plants. The herbicidal auxins are, of course, a case in point. Another interesting kind of growth inhibitor, Amo-1618, has recently been widely publicized as a type of anti-gibberellin. When applied to chrysanthemums, Amo-1618 will cause a tremendous diminution of stem elongation without major effect on leaf size, on flowering behavior, or on the size of the flower. Thus it is possible to grow large-flowered chrysanthemum plants of very small total size. This may turn out to be extremely important in greenhouses and in home gardening. Still another substance, called CMU, is a specific photosynthetic poison, and may thus be used to clear an area of all green vegetation, without any damage to animal inhabitants. Certainly, with continued screening of the many new types of organic molecules being synthesized in various laboratories around the world, we may expect to see the emergence of many useful promoters and inhibitors of plant growth in the years to come.

5 Differentiation and Morphogenesis

Probably the major unsolved problem in biology today is the mechanism of differentiation. The problem may be stated in this way: If all the cells of any organism arise by mitotic division of the fertilized egg, all cells should be endowed with an equal genetic potential, as contained in the chromosomes and in the cytoplasmic organelles. Yet we know that the various cells of the multicellular organism do, in fact, become different in appearance, in function, and in basic chemistry. Students of the problem of differentiation are trying to understand both the factors that give rise to differences among genetically similar cells and the precise kinds of chemical and physical changes occurring in the cell that lead ultimately to the assumption of differences in form and structure. So far, biologists have made little progress in this difficult field.

The student of the physiology of higher plants knows how to control certain differentiation phenomena by chemical and physical means. His knowledge, however, is purely empirical. He is like the man who inserts a key into the door and opens it without knowing anything about the intricate mechanism of the lock. Certainly much of the future progress in experimental biology will depend on some detailed understanding of the phenomena that intervene between the application of the specific agent and the appearance of the altered form.

The word *morphogenesis* refers to the study of the origin of morphological characters and form. It includes the problem of differentiation, as well as other aspects of structural development that occur once primary differentiation phenomena have been completed. Many biological systems can be chosen to illustrate the basic problems in morphogenesis. Let us consider, as an example, the apparently simple problem of the origin of two different kinds of poles in the fertilized egg of *Fucus* (Fig. 60). *Fucus,* commonly called the rockweed, is a member of the brown algae and grows in the intertidal rocks along the ocean front. The diploid plant body consists of a basal attachment disc, the *holdfast,* and a flattened branched body called a *thallus.* At appropriate times of the year, spherical structures, the *conceptacles,* are produced at the end of the thallus. These contain the male and female sex organs, called, respectively, *antheridia* and *oogonia.* The single-celled eggs and sperms found in these sex organs are released into the ocean water, where fertilization occurs. The fertilized egg (zygote) is a spherical cell floating free in the water; it is devoid of any surface characteristics that would indicate the existence of morphological differences between one region and another.

Shortly after fertilization, the zygote no longer appears completely symmetrical, and with the first cell division, a permanent distribution of structural and functional roles occurs between a rhizoidal (root-like) pole and a non-rhizoidal pole. The cells produced by the rhizoidal pole are positive in their geotropic responses, growing downward to produce a stalk and holdfast. The non-rhizoidal pole, on the other hand, gives rise to the horizontal flattened thallus we described earlier.

What is the nature of the influence brought to bear on the developing zygote that causes a cleavage plane to be formed in one particular direction and that causes the assumption of particular structural and functional characteristics by the two different poles? Some years ago

Fig. 60. The life cycle of *Fucus,* the rockweed, a member of the brown algae.

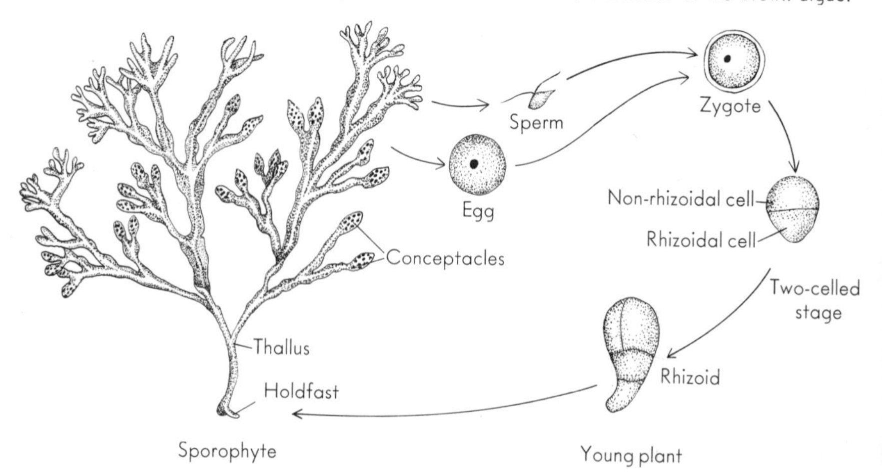

Whitaker and Lowrance found that the rhizoidal pole developed (1) on the side away from visible light, when zygotes are exposed to a unilateral light source, (2) on the side that is in a warmer environment, where a temperature gradient is maintained across the zygote, (3) on the side toward neighboring zygotes, when several are present in a cluster, (4) toward a more acid medium, when a pH gradient is maintained, and (5) at the centrifugal pole of a centrifuged zygote. All these facts are consistent with the establishment across the zygote of some subtle gradient, at first measurable only in physical and chemical terms but later manifested as gross morphological differences over the entire organism.

Let us take another example. We have previously discussed the classical work with the *Avena* (oat) coleoptile, especially as it was related to the study of auxin physiology. You will recall that if a block of agar containing auxin is placed at the upper end of a cut coleoptile, the auxin is rapidly translocated downward and influences the growth of cells below the cut. If, however, a cylinder of a coleoptile is removed and the agar block containing the auxin is placed at the morphological base of this cylinder, there is no movement of auxin into the tissue (Fig. 52). This phenomenon of *polarity* is not restricted to *Avena*, but is found in a great many plant tissues. It obviously means there is some kind of a difference between the morphological apex and base of each individual cell in the *Avena* coleoptile. But this difference is not visible in any way to the microscope, nor can it yet be demonstrated by any chemical means. Whatever the basic explanation of polarity is, it probably bears on other problems in which invisible gradients are established, as in the *Fucus* egg.

Several workers have suggested that the difference between the apex and the base of a coleoptile may be represented by an electrical potential. In fact, attempts to measure electrical potentials in various plant organs have been fairly successful. For example, the tip of the coleoptile has been found to be electrically negative with respect to the base. Yet this is not a satisfactory final answer to the problem of polarity and differentiation, because the difference in electrical potential is itself the result of some other phenomenon that lies at the base of differentiation processes.

In the remainder of this chapter, we shall consider some of the more obvious and vexing problems of differentiation and morphogenesis in the higher green plants and the techniques being used to attack them. Among the more instructive of these are the difference between root and shoot, the difference between vegetative and reproductive primordia, and the difference between normal and atypical cells of various kinds.

Organ and Tissue Culture in the Study of Morphogenesis

As long as plant cells are developing within the framework of organized plant tissue, it is difficult to study the factors that influence the course of their differentiation. A frequently employed technique, therefore, is to remove from the plant a group of cells constituting either an entire organ or a particular kind of tissue. These cells can then be transferred, under bacteria-free conditions, to a nutrient medium in an appropriate container, where their morphogenetic pattern may be studied under a wide variety of deliberately varied conditions. These techniques are referred to collectively as *tissue culture*. As an example of this general approach to problems of plant morphogenesis, let us consider the excised root (Fig. 61).

In the intact plant, the only nutritional requirements that can be defined for the root are those common to all plant cells, like mineral nutrients and some source of energy such as the carbohydrate, sucrose. If, however, a one-centimeter apical zone of a previously sterilized seedling root of, for example, a pea plant is excised and transferred to a nutrient solution containing only sucrose and mineral salts, the root will

Fig. 61. Techniques employed in the aseptic culture of excised root tips.

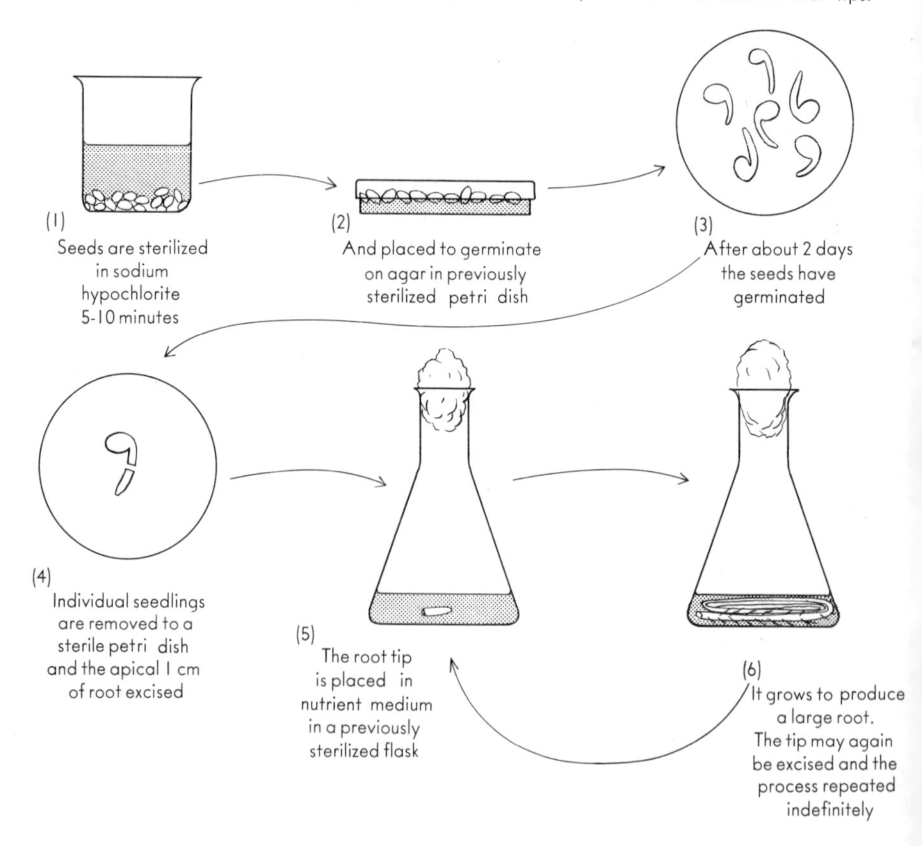

eventually cease to grow. In the late 1930's when this experiment was first done, it was discovered that yeast extract added to the basic salt and sugar nutrient medium would supply the additional materials required in trace quantities for the continued and potentially limitless growth of these roots. As the components of yeast extracts became known, the extra nutritional requirements were discovered to be the vitamins thiamin (B_1) and nicotinic acid. If these are supplied to the nutrient medium in pure crystalline form at a concentration of about 0.1 to 0.5 mg/L, the growth of the excised root continues indefinitely.

Thus it appears that the root is biochemically differentiated from the stem in not being able to synthesize the vitamins thiamin and nicotinic acid. Since all cells need these substances (they are coenzymes for important enzymes), we can assume that the leaves and stems supply to the roots all the thiamin and nicotinic acid the latter need. This theory can be tested by blocking the transport of these materials, through the technique of steam-girdling the phloem of the stem. The transport of materials in the xylem, mediated entirely by nonliving cells, is not interrupted by such treatment, but the transport of materials in the phloem is completely halted by the destruction of the living sieve-tube elements. In such a steam-girdled stem, the vitamins thiamin and nicotinic acid accumulate above the girdle and become depleted below the girdle. We may conclude, therefore, that in the course of normal plant nutrition these vitamins are synthesized in the aerial parts of plants and are transported down to the roots where they are used for normal metabolism and growth. The fact that these substances are made in microquantities in one part of the plant and are transported to another part of the plant where they are used in growth and development qualifies them as hormones.

In general, the excised tip of a root, when grown in a properly supplemented medium, will continue indefinitely to produce new organized root tissue. With some roots, branch roots may be formed in culture in the usual way, arising in the pericycle region some distance behind the apex. The secondary roots may grow as vigorously as the primary root, although in some instances they fail to grow when subcultured on the same medium that is adequate for the primary root. The application of auxin to root cultures tends to increase the number of branch roots formed. The root tip appears to exert some inhibitory influence, on branch-root formation, since removal of the tip generally leads to an increase in the total number of roots formed. Branch-root initiation and root growth generally proceed best in darkness and are inhibited by very low total irradiances of visible light.

In certain roots, such as those of morning-glory, *Convolvulus*, adventitious buds can be initiated under specific conditions. In general, these conditions involve an absence of auxin, the presence of light, and, in some instances, the addition of substances analogous to the purines found in nucleic acid. Kinetin, discussed above, is quite effective in this process. Differentiation into root tissue, therefore, does not, at least in the case of *Convolvulus,* involve an irreversible loss of the potential to form stem and leaf tissue. It is generally true however, that roots have much lower morphogenetic versatility than do stem tissues.

CULTURE OF EXCISED STEMS

The apex of a stem, like the apex of a root, may be removed from the plant and placed on a synthetic nutrient medium, where it can be made to grow with ease. Stem tissues are remarkably autotrophic, requiring only a source of energy in the form of the sucrose molecule and the usual mineral nutrients already discussed. When cultured in the light, stems generally become photosynthetic and are then completely independent of exogenous sucrose. Almost all stem tissues, without any special urging, will form root primordia spontaneously, especially if cultured in the dark. These root primordia will then grow out to form large functional roots, and, in effect, the stem tip will have regenerated an entire plant. This obviously means that the stem tissue has the morphogenetic potential to produce roots, just as some root tissue has the morphogenetic capacity to form stems. In fact, there are only two types of stems known that can be grown in culture potentially indefinitely without the formation of root tissue. These are the stem tips of asparagus and of dodder. Asparagus stem tips can be grown completely autotrophically in the light, since they are capable of photosynthesis (Fig. 62). In bright light, they rarely give rise to anything but stem tissue plus the slightly altered "cladophylls" or flattened stem parts resembling leaves. If such stems are transferred to darkness and supplied with sucrose, they grow very rapidly, and root primordia occasionally form. Root initiation may be significantly increased by the application of auxins, which again are especially effective in this process if the stems are kept in a dark place.

Dodder stems are unique in that they not only grow without the formation of roots, but also give rise to normal flowers in culture. This is one of the very unusual instances in which an excised, cultured stem tip is able to realize its morphogenetic potential for the production of floral organs.

In general, then, stems can be grown as isolated systems only in unusual instances, for their normal tendency is to produce root primordia spontaneously in their basal regions. This phenomenon is probably due to the basal accumulation of auxin in the stem as a result of downward transport of this substance. The tendency to form roots is somehow inhibited by light and favored by the application of exogenous auxins.

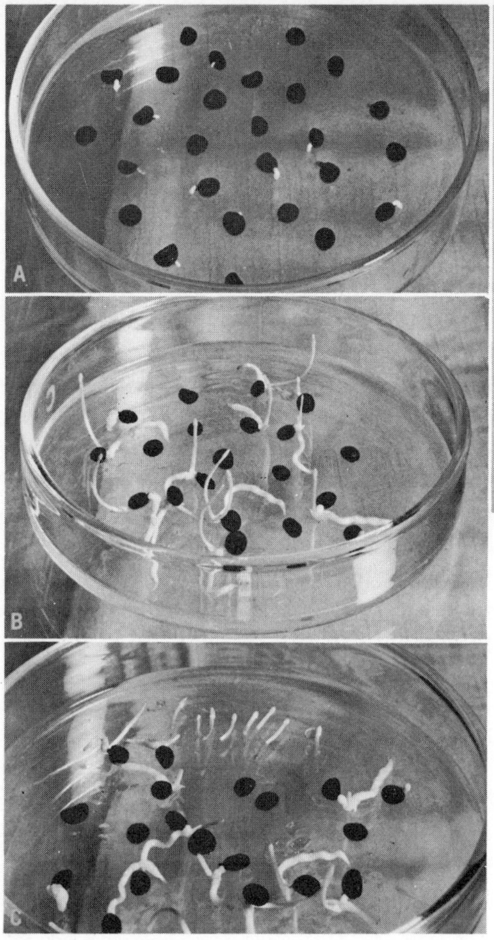

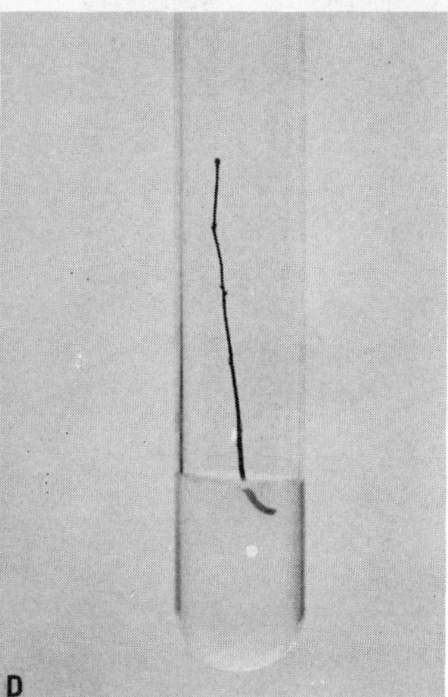

Fig. 62. Steps in the culture of excised asparagus stem tips. (A) Sterilized asparagus seeds on agar; several are beginning to germinate. (B) Germinated seeds with erect stems and prostrate roots. (C) Stem tips excised and laid out ready for transfer. (D) A stem tip that has grown in the light for two weeks. (From A. W. Galston, "On the Physiology of Root Initiation in Excised Asparagus Stem Tips," *American Journal of Botany*, 35 [1948], 281–287.)

LEAF CULTURE

It is sometimes possible to excise a young leaf from the parent plant and grow it successfully in culture to the mature stage. The culture medium in such experiments must contain an energy source, the usual mineral nutrients, and usually specific organic materials (Fig. 63). The substance, kinetin, discussed above, can greatly increase the growth of excised leaf discs placed on synthetic media. In this action, kinetin acts together with light, which also promotes leaf growth. Various people have attempted to analyze the mode of action of kinetin in producing this effect and have been able to show that it favors the accumulation of nitrogenous substances in the leaf and also retards protein loss. For example, if an excised tobacco leaf is cultured in water, it soon becomes yellow and dies. If kinetin is placed on one half of the leaf, that half will continue to live long after the control half has died. If kinetin is placed at one location on the leaf, and carbon-14-labeled amino acids are placed on the other

Fig. 63. The growth in culture of excised fern leaves. (Top) Young frond. (Bottom) Somewhat older fronds. (Courtesy T. A. Steeves and I. M. Sussex, "Studies on the Development of Excised Leaves in Sterile Culture," *The American Journal of Botany*, October, 1957.)

side of the leaf, these acids will migrate to and accumulate at the locus of kinetin application. This probably results from an effect of kinetin on nucleic acid metabolism that ultimately gets translated into the synthesis of proteins from amino acids.

Several decades ago evidence was obtained for the existence, in cotyledons of peas, of substances that would specifically promote the expansion of excised leaf discs. An analysis of the effective extracts revealed that the substances adenine and hypoxanthine, which are purines, could replace, at least in part, the effect of the mixture. This finding would be consistent with our more recent knowledge that kinetin is effective in promoting leaf growth and would tend to imply that the growth of the green leaf is intimately connected with nucleic acid metabolism.

In some plants, such as the African violet, new organs can be regenerated from the base of the petiole of the excised leaf. The first step in such a process is usually the initiation of root primordia, which grow out into roots, followed by a growth of adventitious buds to form stem material. Thus an entire plant can be regenerated from a single excised leaf, showing once again that differentiation into a particular organ does not involve an irreversible loss of potential to form other kinds of structures.

CALLUS CULTURE

In many respects, the most interesting type of culture experiments with plant cells involves the culture of *callus* tissues (Fig. 64). Such calluses, which may be derived by the excision of fleshy tissues of various plants such as potato tubers and carrot roots, may be grown potentially indefinitely on very simple media. These media need contain

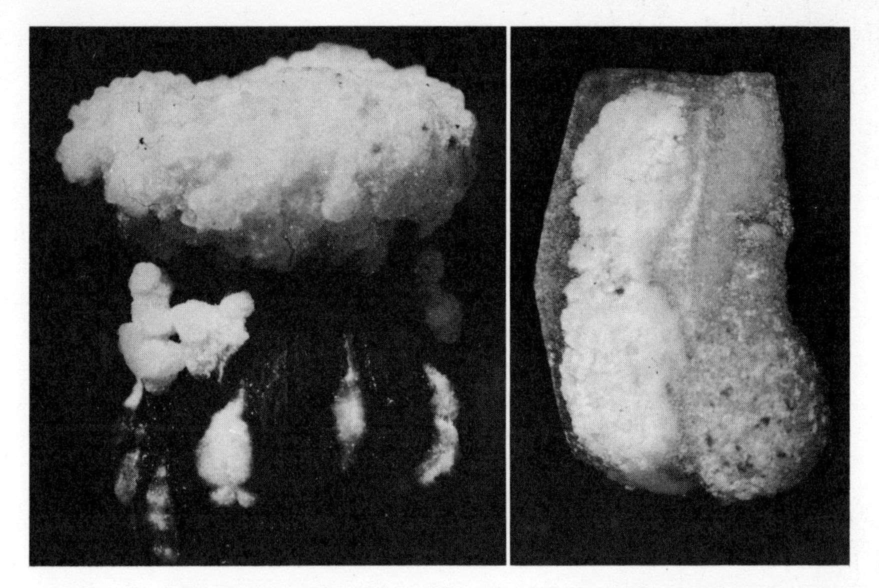

Fig. 64. (Left) A callus culture of grape stem after 60 days of growth. (Courtesy R. J. Gautheret.) (Right) A piece of carrot root that has been allowed to develop on synthetic nutrient medium. The light areas represent new groups of cells that differ in form and chemistry from the cells making up the bulk of the root. (Photo by A. W. Galston.)

only sugar, mineral salts, and a reasonable concentration of some auxin. The cells of such callus cultures are large in size, are highly vacuolated, and are capable of many morphogenetic alterations. For example, deep within the mass of such tissue, isolated whorls of lignified cells resembling the tracheids of a normal plant are often detected. Although these cells are rather normal in their morphology, they clearly do not function in transport. The fact that they always arise deep within the fleshy part of the tissue implies that there is something about an interior location that favors the differentiation of lignified elements.

It has also been shown in a great variety of tissue cultures that auxin favors the initiation and development of lignified cells. This was discovered by means of an experiment in which a normal bud of lilac was grafted onto an undifferentiated callus of lilac tissue. The bud influenced the callus tissue to develop an organized strand of vascular tissue that ultimately became linked with the vascular tissue of the bud (Fig. 65). This occurred because the lilac bud produces a substance that passes the wounded area readily. If the bud is replaced by a high concentration of auxin in agar, xylem tissue again develops in the callus, but is not as well oriented as when normal tissue of the bud is present. In this xylem-induction phenomenon, an adequate level of sugar also appears to be very important. We may thus conclude that buds exert their organizing influence on callus tissue by supplying auxin, together with organic substrate, to the area in which xylem cells are developing.

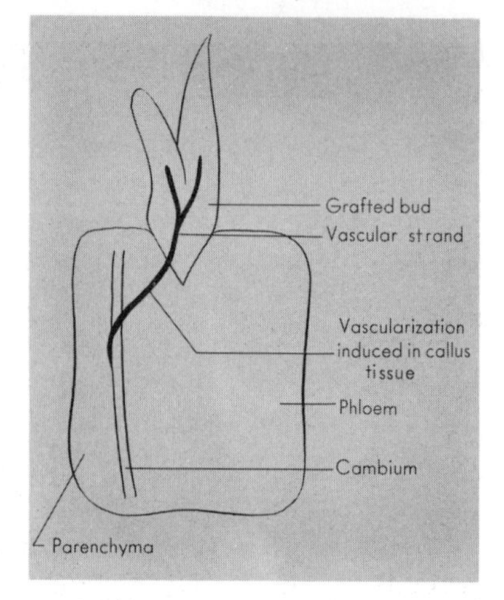

Fig. 65. The effect of a bud on vascular differentiation in a callus to which it had been grafted. (After G. Camus.)

The loss of differentiated cell types in the callus is not a sign of an irrevocable loss of ability to organize normal tissue patterns, for, under appropriate conditions, the normal patterns can reappear. For example, in the presence of a high concentration of auxin, callus tissue frequently gives rise to organized roots. In other instances, such as in tobacco pith callus, the application of kinetin or of various purine materials will result in bud initiation. Workers with this tissue have proposed that the morphogenetic pathway followed by a mass of callus tissue may be the result of the simple ratio of two kinds of substances, the auxin type and the kinetin type. If the auxin type predominates, roots will tend to be initiated; if the kinetin type predominates, buds will tend to be initiated; if the two are in balance, growth will tend to continue as a callus. Although these generalizations seem to hold well for tobacco pith callus, they do not appear to pertain to other callus tissues that have been cultured.

Sometimes callus-like growths develop normally on plants, and, when subcultured, they continue their atypical pattern of growth. A familiar example is the leguminous root nodule (discussed in Chapter 3) produced as a result of invasion of a root hair by the bacterium, *Rhizobium.* It is now well known that the bacterium penetrates near the tip of the root hair and produces a long filamentous "infection thread" which somehow alters the metabolism and growth potential of the cells in the region of the infection in such a way that normal morphogenetic patterns are obscured, and the knotty masses of nodules are formed instead. The new tissue is different from either bacterium or host in several important respects. The most important, of course, is its ability to absorb atmospheric nitrogen and transform this into ammonia, which is then used by cells to make amino acids, proteins, and other nitrogenous materials.

As mentioned previously, such nodules contain a red pigment that is in many respects similar to the hemoglobin of red blood cells. This pigment, named *leghemoglobin,* is assumed to be important in biological nitrogen fixation by the nodule, because the most active nodules tend to have the highest concentration of leghemoglobin and the least active tend to be low in leghemoglobin. Here, then, is a case where the cell's morphogenetic and biochemical potential is altered by new biological materials entering the cell through a bacterium. Nodules may be initiated on leguminous roots grown in pure culture by the introduction of the appropriate *Rhizobium*. If the nodule tissue is excised, it grows as a nodule, not as the normal type of root it was derived from. This may represent the permanent loss of morphogenetic potential.

A more striking example of a permanent alteration of the morphogenetic potential of the cell is caused by the inoculation into a wounded cell of the organism *Agrobacterium tumefaciens*. If a virulent culture of the bacterium is injected into a susceptible cell, such as that of a sunflower, a large tumor-like mass will be produced after some time (Fig. 66). This tumor competes effectively with the rest of the plant for growth-promoting materials, and, as a result, normal vegetative growth may be retarded tremendously. The cells of this primary tumor contain the bacterium which, although greatly altered in form, can still be recovered and shown to consist of the same virulent bacteria that were originally introduced into the sunflower.

In the development of this crown gall disease, secondary tumors may arise at some distance from the initial site of inoculation. Such secondary tumors, if excised and transferred to a nutrient medium for culture, will also continue vigorous growth as a tumor. All attempts to recover the bacterium from such secondary tumors, however, have been

Fig. 66. A crown gall tumor produced on a geranium stem by inoculation with *Agrobacterium tumefaciens*. (Courtesy G. Morel and R. J. Gautheret.)

unsuccessful, indicating that the bacterium is not itself the infective agent, but is merely the package by means of which some "tumor-inducing principle" is introduced into the plant cell. This tumor-inducing principle, once released from the bacterium, can be transported about the plant independently of bacterium.

Crown gall cells differ from normal callus cells in several interesting respects. First, they are completely independent of exogenous auxins. We have already mentioned that the normal callus requires a source of auxin for continued growth. Crown gall cells, however, are actually inhibited by the addition of auxin. It can be shown, furthermore, that the crown gall cells grown without auxin are very high in auxin content, and we naturally conclude that they have somehow acquired an ability to produce enough auxin for their growth needs. This, of course, may have something to do with their tumorous nature. It may explain why they are able to grow competitively with other cells, which are dependent on the normal auxin supply from the apex of the plant. But the difference between normal and crown gall tumor cells does not lie solely in their auxin metabolism, for the application of auxin to a normal cell does not convert it into a crown gall cell. Indeed, the crown gall cell is capable of transmitting a tumor-inducing principle to receptor cells to which it is grafted. A successful graft will result ultimately in the production of still other tumors on the receptor plant. In many respects, this tumor-inducing principle resembles a virus that can be transmitted from plant to plant through living cells, but such a material has not yet been isolated from crown gall cells.

Although crown gall tissue is not promoted in its growth by the addition of exogenous auxins, it is greatly promoted in over-all growth rate by the application of an unknown complex of substances in coconut milk or in the liquid endosperms of other young fruits. The nature of the effective materials in coconut milk is unknown; it clearly is not auxin.

The crown gall disease of plants is in many respects an analog of animal cancer. In fact, many investigations of atypical and cancerous growth have been and are still being performed with crown gall tissue. If the chemical nature of the tumor-inducing principle becomes known, this information may shed some light on the nature of substances causing animal cancer. In addition, animal cancer cells, like the crown gall cells, may possibly become autonomous with respect to a particular growth factor such as a growth hormone. Both plant and animal tumor cells are quite different from normal cells in their morphogenetic potential. For example, unlike callus tissues, crown gall cells have never been observed to organize root primordia or bud primordia. They also do not normally give rise to the tracheids and other organized elements deep within the callus mass. The introduction of the tumor-inducing principle into the crown gall cell seems forever to have blocked the possibility of normal development. This, in essence, may be the basis of tumor action.

THE CULTURE OF SINGLE CELLS

The problems of plant morphogenesis are highlighted by a consideration of the culture of isolated plant cells. If a single cell of known type can be isolated and grown to a mass of tissue, and if such masses of tissue can be tested for their ability to organize particular plant organs, questions can be framed and answered that give us a much greater appreciation of the nature of differentiation and morphogenesis (Fig. 67).

Fig. 67. (Top) Method of growing single cells in culture. (Bottom) Several typical isolated cells. (B1, B4) Living cells from a tobacco pith culture. (B2, B3) Tracheid-like cells from a marigold culture. The walls are stained with phloroglucinol to show thickenings. Single isolated cells of the types shown in B1 and even in B4 were used successfully in the production of cultures. (From W. H. Muir, A. C. Hildebrandt, and A. J. Riker, "The Preparation, Isolation, and Growth in Culture of Single Cells from Higher Plants," in *American Journal of Botany*, 45 [1958], 589-597.)

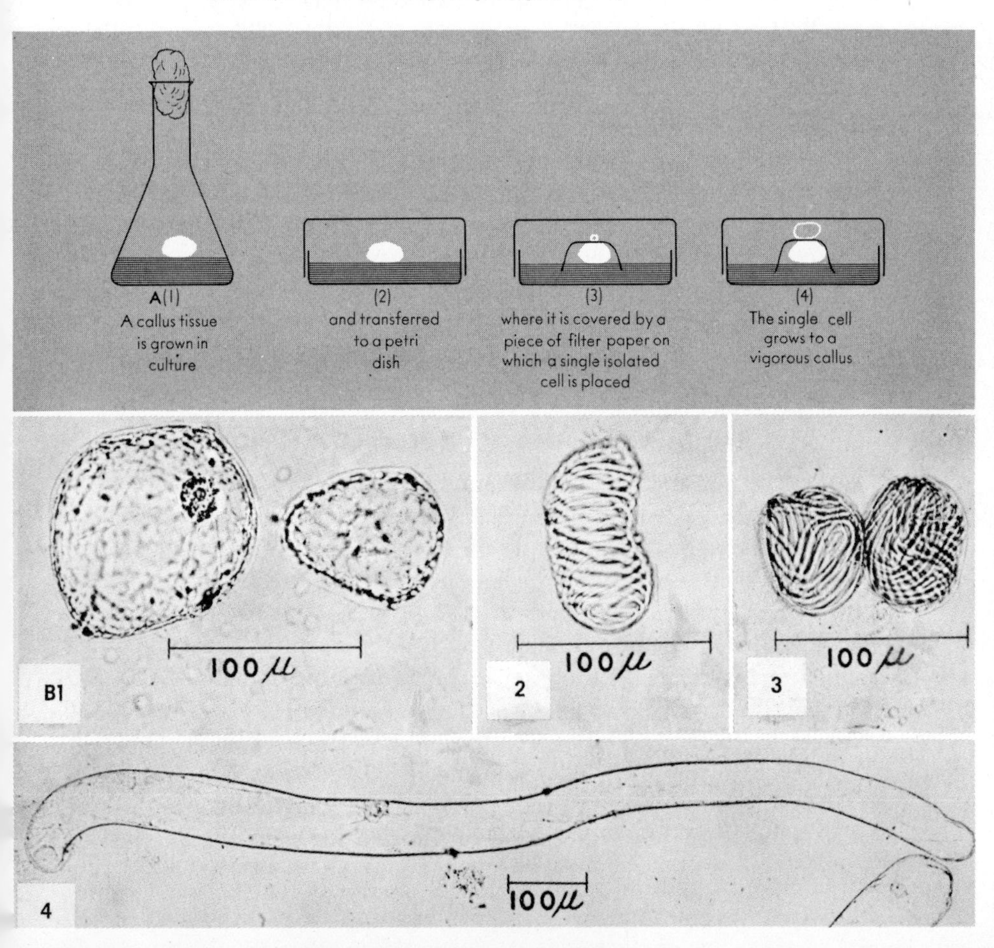

For example, does each individual cell of the mature plant have the morphogenetic potential to restore the intact plant? This question seems to be answered in the affirmative by a series of experiments conducted by various researchers in the last three years. The isolation of an individual plant cell is difficult because of the tendency of plant cells to cohere to form masses of cells. The most convenient way to get isolated cells is by mechanical agitation, i.e., by the physical separation of one cell from its neighbor. Another useful technique is to use chemical agents that bind the element, calcium, which is important in the formation of calcium pectate, the intercellular cement of plants. A third method is to employ the enzyme, pectinase, which dissolves the intercellular cement and separates the free cells.

The single cell, however obtained, shows great reluctance to grow when placed by itself into a medium that normally suffices for the growth of a larger mass of cells, apparently because it loses to the medium large quantities of materials that are required for growth. The easiest way to overcome this difficulty is to provide the cell with "nurse tissue" in the form of masses of cells of its own type that are kept nearby but are still physically isolated from the cell in question. This may be done either by placing a single cell on a filter paper that sits on top of the "nurse tissue," or by placing the single cell on an agar medium close to, but not touching mature tissue.

Such single cells may form multicellular masses, which then give rise to different cell types and ultimately to all the differentiated organs of the intact plant. This cycle has been carried out most successfully with carrot cells, but there are now signs that it may be a general phenomenon. If these results are generally true, it indicates that in the process of differentiation there is no irreversible loss of morphogenetic potential by any of the cells of the body of the plant, or at least by those that can be isolated and made to grow in culture.

The Differentiation of Reproductive Organs

One of the most dramatic examples of the alteration of form of a plant occurs at the time that the plant makes the transition from vegetation to reproduction. For example, in many angiosperms, the plant produces root, stems, and leaves for a long period of time. Then, at some point in its life history, it ceases vigorous vegetative growth and begins a series of transformations leading ultimately to the production of the reproductive organ, the flower, containing sepals, petals, stamens, and carpels. In the stamen and in the carpel will occur the important process of meiosis, which leads to the production of the haploid stage (gametophyte) of the life cycle of the plant. The male and female gametophytes ultimately give rise to the sex cells, i.e., the sperms of the pollen tube and the egg of the embryo sac. With the union of the two sex cells, the $2n$ (sporophyte) generation is restored, and the life cycle is therefore complete.

Within the last several decades, much knowledge has been accumulated about the environmental factors that control the onset of reproduction and the nature of the intervening chemical changes that lead to this altered form. It will be instructive, in our consideration of morphogenesis, to delve into this important process. Before going to the angiosperms, let us consider a simpler type of system, the prothallus of a fern (Fig. 68). The fern prothallus is the haploid (gametophyte) generation, i.e., each of the cells has only one-half the normal chromosome complement of the mature fern plant. The gametophyte is formed by repeated mitotic division from a single-celled haploid spore. Ultimately, the prothallus consists of a single layer of cells, rather heart-shaped in appearance, bearing rhizoids, or root-like structures, growing downward into the substrate. At some stage in its life cycle, the prothallus will initiate both antheridia, the male sex organs, and archegonia, the female sex organs. Within each antheridium will be borne a group of sperms, and within each archegonium will be borne a single egg.

The question may now be posed: What governs the differentiation of the cells of the prothallus into each of the sex organs? The answer seems to lie in the production of specific substances that control differentiation. For example, if spores of the sensitive fern, *Onoclea*, are placed

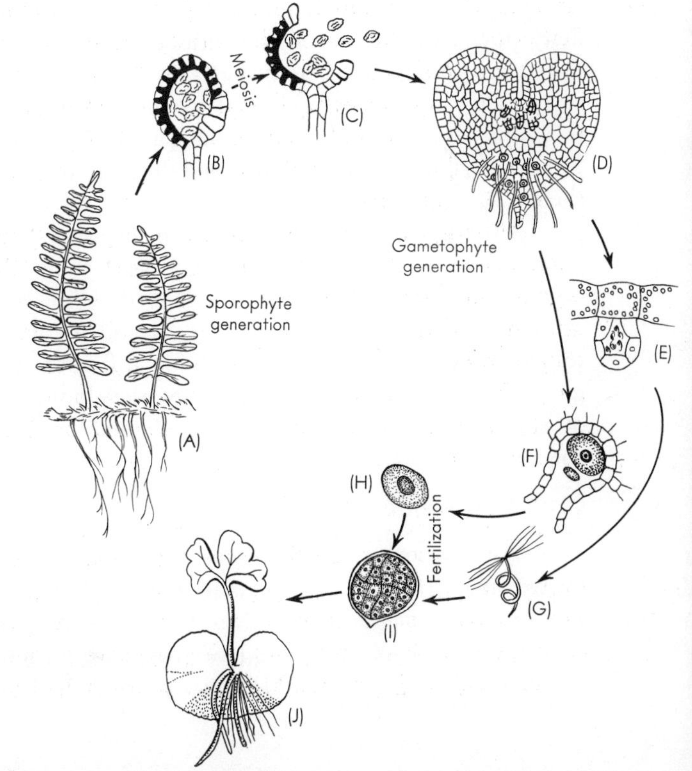

Fig. 68. The life cycle of a fern. (From Carl Swanson, *The Cell*. Englewood Cliffs, N. J.: Prentice-Hall, 1960.)

singly in test tubes containing a synthetic medium adequate for growth, the prothallia will develop normally, except that they will never give rise to antheridia. If, however, they are grown in a medium containing many other prothallia or the culture filtrate from older prothallia of *Onoclea,* or of other related ferns such as *Pteridium,* antheridia are initiated promptly. Simple dilution experiments show that in the old *Pteridium* medium a substance exists that specifically causes the differentiation of antheridia on the prothallus, without having any marked effect on its growth rate. If large quantities of this material are added to the culture medium, almost every cell of the mature prothallus will give rise to antheridia.

The interesting fact that emerges from a detailed study of this phenomenon is that by the time a prothallus has gotten old enough to produce the antheridium-inducing factor, it is too old to respond to it. Therefore, a single prothallus cannot induce itself to form antheridia. This implies also that the production of antheridia in nature depends on the simultaneous presence of many prothallia of heterogeneous ages. The old ones produce the specific factor that induces antheridial formation in the younger prothallia, but themselves go on to form archegonia directly, without prior antheridial formation. Thus the successful differentiation of the sex organs, which in turn eventually leads to completion of the sexual reproduction of the plant, is dependent on "togetherness." An analysis of various families of ferns reveals that there is some specificity in the production of this substance. By and large, all members of the family *Polypodiaceae* respond to the *Pteridium* factor, but do not respond to substances produced by fern prothallia of other families. In the same way, non-Polypodiaceous gametophytes will not respond to the *Pteridium* factor, but will respond instead to substances produced by their own prothallia.

Although the analysis has not been carried quite this far, there is good evidence that in certain algae and fungi the production of sex organs, the maturation of sex organs, the release of gametes, and the attraction of the sperm to the egg are all mediated by particular substances produced by the organisms at different stages of their development. This reinforces the view that development consists of a series of stages, each of which may be elicited or triggered off through a specific chemical event.

PHOTOPERIOD

We now know that the flowering of many angiosperms is controlled by two major factors of the environment, photoperiod and temperature. The discovery of photoperiodism by Garner and Allard in 1920 was the accidental result of an attempt to propagate a mutant type of large-leaf tobacco, called Maryland Mammoth, which had arisen by chance as a single individual in a field of other tobacco plants. As the season progressed, the original type flowered profusely, but the Maryland Mammoth did not. Wishing to obtain seeds of this valuable new type and fearful that the plant might not flower before the autumn frost, these investigators removed the plant from the field and transferred it to the greenhouse (Fig. 69). Despite every urging, however, the plant steadfastly refused to initiate floral organs until approximately mid-December, many months after the normal plants had completed seed production successfully. When the seeds of the self-pollinated Maryland Mammoth type were planted in the field the next year, this behavior pattern was repeated, i.e., the plants grew vigorously in the field, failed to flower together with the original type, but did flower when removed to the greenhouse and maintained until about Christmas time.

Fig. 69. Maryland Mammoth tobacco plants grown under short-day (left) and long-day (right) conditions. (From A. E. Murneek and R. O. Whyte, *Vernalization and Photoperiodism*. Copyright 1948 The Ronald Press Company. Photo by Garner and Allard.)

An analysis of the various factors that could possibly be responsible for this behavior finally led Garner and Allard to the inevitable conclusion that the plant was flowering only during the very short days characteristic of the Northern Hemisphere at Christmas time. They discovered that flower initiation could be induced at will by transferring Maryland Mammoth plants to special chambers in which the length of day could be shortened to about 9 hours. They named this response of plants to length of day *photoperiodism*. Maryland Mammoth tobacco, which flowers only if the day length is reduced below a certain critical value, was called a *short-day plant*. Other plants of this type are soybeans and chrysanthemums. Another type of plant is the *long-day* type, such as

Fig. 70. (Upper left) A cocklebur plant, a wooden holder, and an opaque paper container. (Lower left) The cocklebur plant inserted into the paper container. The enclosed upper portion receives short-day treatment; the exposed bottom portion continues to receive long-day treatment. (Right) Results. Both the enclosed, induced portion and the exposed, non-induced portion have flowered and formed fruits. Some mobile stimulus must therefore have moved from the induced to the non-induced portion of the plant. (Courtesy K. C. Hamner and J. Bonner, "Photoperiodism in Relation to Hormones as Factors in Floral Initiation and Development," *Botanical Gazette,* 100 [1938] 388–431, University of Chicago Press.)

spinach and certain cereals, in which flowering occurs only if the day length exceeds a certain critical value. Finally, there is a class of plants called *day-neutral plants,* in which photoperiod does not exert a major effect on the time of flower initiation. An example of this type is the tomato plant, in which floral primordia are initiated at particular nodes when the plant has obtained a particular size. This situation is not amenable to control by photoperiod, although the tomato may be induced to premature flowering by certain synthetic chemicals that are related to auxins.

In the years since 1920, considerable work has been done to determine how photoperiod exerts its effect on the flowering of the plant. It has been unequivocally demonstrated that the leaf is the receptor for the photoperiodic stimulus. For example, if a single leaf of a Maryland Mammoth tobacco plant is enclosed in a black bag and given the appropriate short photoperiod required for floral initiation, the terminal bud some distance away from the leaf will initiate floral primordia. This separation in space of the photoperiodic receptor (leaf) from the region of response (bud) makes it necessary to postulate some connecting link between the two (Fig. 70). Since it can also be demonstrated that a plant exposed to a short day can transmit its florally-initiated state to a receptor plant maintained vegetatively on a long day, it is clear that some hormonal substance, tentatively named *florigen,* is produced in the photoperiodically-stimulated leaf. This stimulus can be transmitted from a donor plant to a receptor, provided there is tissue union between the two grafted plants. The transmission of the stimulus is prevented by steam-girdling of the petiole or by other means of interrupting phloem transport. These facts clearly imply the production in the photo-induced leaf of a yet unknown substance, which is transported via phloem to the terminal bud, where it influences the meristem to favor the production of floral primordia rather than vegetative organs.

Recently it has become clear that many long-day plants may, if supplied with gibberellin, initiate floral primordia under an otherwise unfavorable photoperiod (Fig. 57). In this group of plants, there appears to be some connection between gibberellin and florigen. Gibberellin, however, does not promote flowering in short-day plants. Since there is very good evidence from grafting experiments that the florigen of long-day and of short-day plants is at least functionally equivalent, if not chemically identical, the exact nature of the relation between gibberellin and florigen is not clear. In still other plants, such as the pineapple and the litchi, floral organs can be initiated by the application of certain synthetic auxins. In the series of events leading to the production of reproductive organs, therefore, various substances may well become limiting in different plants. This substance may be gibberellin in some plants, auxin in other plants, and, perhaps additional, possibly unknown substances, in still other plants.

One interesting generalization that has developed out of the study of these photoperiodic phenomena is that most plants respond not to the length of the light period but to the length of uninterrupted darkness. Thus a so-called short-day plant is really a "long-night" plant that requires an uninterrupted dark period of a certain minimal duration for the initiation of its floral primordia. In the same way, a long-day plant is in reality a "short-night" plant, that is, a plant that will flower only if the night period is not longer than a certain critical maximum.

The effective period of darkness can be interrupted by the simple expedient of making it a bit too short (even a few minutes will do), or by interposing a brief flash of light in the middle of the dark period (Fig. 71). For example, in the short-day plant, *Xanthium,* flowering will occur in a regime of 15 hours of light and 9 hours of dark. If the 9-hour dark period is shortened appreciably, to about 8½ hours, flowering will not occur. But one single cycle of exposure to 15 hours of light and 9 hours of dark will suffice to induce the plant to initiate floral primordia, even though photoperiods unfavorable for floral initiation are immediately re-imposed. Such a phenomenon is referred to as *photoperiodic induction.* If the single long dark period of 9 hours is interrupted at its midpoint by a flash of light, the plant will not flower. Certain chemical processes that are very sensitive to minute quantities of radiant energy must thus be proceeding in darkness in the leaf. If a quantum of absorbed light impinges on the course of these reactions, the entire sequence of events is wrecked and the plant must start over again. With the long-day plants, the situation is just the reverse, i.e., the interruption of an unfavorably long dark period by a brief flash of light will lead to floral initiation. These two types of plants seem to possess the same kind of photoperiodic mechanism, but they somehow work in reverse fashion.

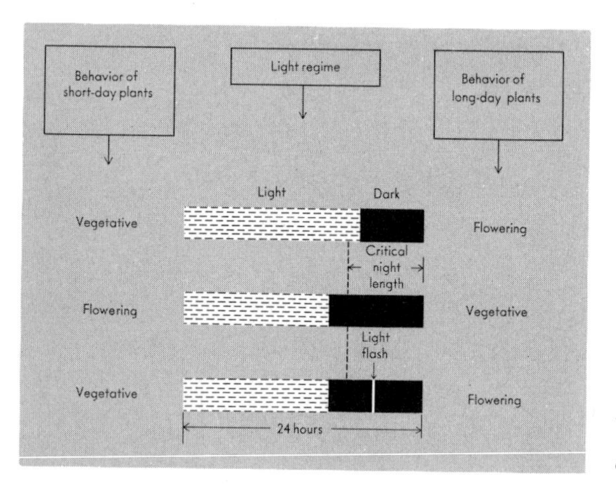

Fig. 71. The effect of a light-flash interruption of the dark period on flowering in short-day and long-day plants.

By the use of the action spectrum technique described earlier, it is possible to obtain some notion of the wavelengths of light that are effective in interrupting or promoting flowering when administered in short flashes. We should mention here that the same kind of light that inhibits the flowering of short-day plants will promote the flowering of long-day plants. Recent experiments with monochromatic light have revealed that many types of plants respond best to red light in the region near 660 mμ. It has also been surprising to find that the effect of red light may be instantaneously and completely negated by the subsequent application of what is called near-infrared light or "far-red" light in the region of 730 mμ (Fig. 72). These experiments indicate that there is a pigment in plants present in two forms, a red-absorbing form and a far-red-absorbing form. In the short-day plant, the absorption of red light in the middle of the long inductive dark period leads to a negation of flowering, while the absorption of far-red after the red leads to the re-promotion of flowering.

The control of flowering thus appears to be a resultant of the ratio of the two forms of this pigment in the plant. So far, we do not know exactly what this pigment is, although extracts of it that show reversible spectral changes on irradiation have been obtained from several plant

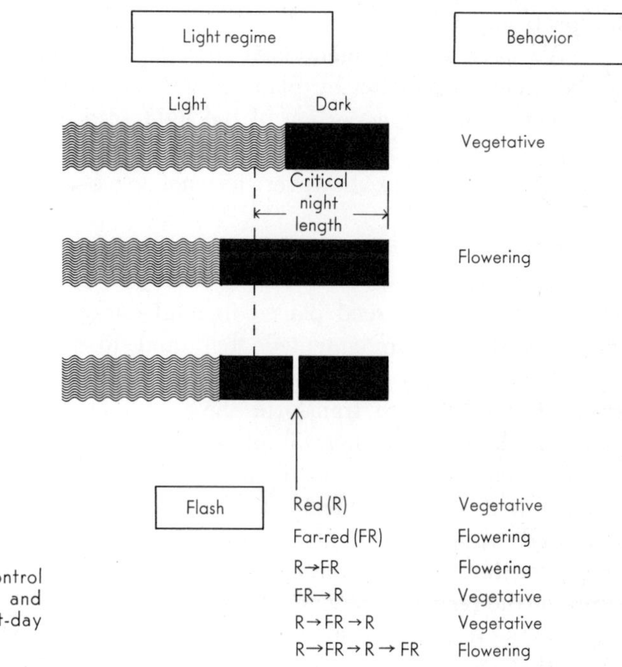

Fig. 72. Reversible control of flowering by red and far-red light in a short-day plant.

tissues. For example, the administration of red light has resulted in a decreased absorption of red near 660 mμ and an increased absorption of far-red near 730 mμ, while administration of far-red has resulted in the reverse changes. Our best guess is that this light treatment is causing a reversible chemical change in the effective pigment molecule and that these changes are determining the course of development of the plant. The yet unknown pigment has recently been named *phytochrome.*

The discovery of the existence of this reversible photoreaction governing flowering has clarified several other perplexing problems in plant physiology. For example, it is well known that the germination of many seeds is affected greatly by light. Seeds of the Grand Rapids variety of lettuce will not germinate at all when placed in darkness on moist filter paper at room temperature, but the administration of minute quantities of red light will result in prompt germination. If the red-light-treated seeds are promptly irradiated with far-red, the effect of the red is completely canceled out and the seeds remain dormant. Here again, the growth of the plant is apparently controlled by a two-way switch mediated by the unknown red-far-red pigment. With other seeds, such as the California poppy, germination is inhibited by light. Here, too, the red-far-red system is involved; red light promotes and far-red inhibits germination. The inhibition, rather than promotion by white light in these seeds, is the result of a reverse differential sensitivity to the two regions of the spectrum.

We have already mentioned that the germination of certain seeds can be greatly promoted by plant growth substances such as gibberellin and kinetin. Indeed, a portion of the light response of these seeds may be interpreted in terms of the alternation of the levels of such substances within the seeds. This, however, has not yet been experimentally demonstrated.

The same controlling red-far-red morphogenetic photoreaction can be seen to operate in the case of *etiolation* (growth in darkness) of stems and leaves. A seed placed in total darkness gives rise to a very long and slender unpigmented stem and to scale leaves that never expand very greatly (Fig. 73). If we analyze the visible spectrum in terms of its ability to transform the etiolated plant into a normal plant, we find that red light is again most effective and that its effect can be prevented by far-red light administered after the red. However, the exact nature of the response depends on the tissue. If the red light is given to stem tissue, that stem will then be greatly inhibited in its growth, but a leaf exposed to the same red light will be promoted in its growth. The responses of both stem and leaf will be completely prevented by the application of far-red light after the red. Therefore, we see that the response to visible radiation is predetermined by the differentiation processes that have given rise to the specific types of cells.

Fig. 73. (Left) Etiolated and (right) light-grown bean plants.

The red-far-red morphogenetic photoreaction described above is certainly of very great importance in the life of the plant, affecting such diverse processes as the germination of seeds, the growth of roots, stems, and leaves, and the initiation of floral primordia. Clearly, increased knowledge of the nature of phytochrome and of its mode of action is greatly needed and must be an important aim of plant physiological research.

MORPHOGENETIC EFFECTS OF TEMPERATURE

In addition to photoperiod, the flowering of plants may be controlled by temperature. It is well known that certain plants exist in annual and biennial strains. The annual strain starts growth in the spring, initiates flowers in the summer, and produces ripe fruit and seeds by the same fall. The biennial variety, on the other hand, produces only vegetative organs during its first growing season. It will not initiate floral primordia until it has first been exposed to prolonged periods of low temperature. Only after such exposure is it able to respond to a proper photoperiodic stimulus. This requirement for low temperatures can be satisfied at any point in the developmental history of the plant after germination. For example, if a biennial seed is permitted to imbibe some water to start germination and is then exposed for about 6 weeks to low temperatures (ca. 2–5°C), it will behave as if it has gone through the cold winter after a year of growth, and will flower in the first season if exposed to the proper photoperiod. This administration of cold treatment to a plant in order to fulfill a specific low temperature requirement for the onset of reproduction is called *vernalization.* Since the difference between the annual and biennial races of most plants seems to reside in a single gene, the low temperature treatment appears to be a substitute, in the biennial form, for some biochemical event produced under genetic control in the annual strain.

Here, again, gibberellin seems to enter the picture, for in several biennials so far studied, the application of gibberellin to an unvernalized plant results in the prompt initiation of floral primordia when that plant is exposed to a favorable photoperiod. Whether cold treatment has anything to do with the synthesis of gibberellin in the plant is not yet clear. At any rate, we know that the vernalization stimulus, like the photoperiodic stimulus, is mobile under certain conditions and can be transported from donor to receptor. The stimulus, therefore, is presumably hormonal in nature, although in the intact plant, where the terminal bud itself is the organ perceiving the cold stimulus, movement need not occur.

Endogenous Rhythms

The phenomena of morphogenesis are much more complicated than we have been able to describe, because the cell itself is more complex than we yet appreciate, and the organism, as an aggregate of complicated cell units, is more complex still. Some of the complexity of the cell can be seen visually, through the electron microscope; some can be detected by physical and chemical means, such as the elucidation of the structure of protein and nucleic acid macromolecules; some can be deduced only from complex behavioral patterns of biological systems that indicate the existence within cells of machinery too complex for us even to describe at present. Such is the case with the endogenous rhythms of plants.

Any event that occurs with a regular periodicity in time or space is said to be periodic or rhythmic. When the periodicity or rhythm persists in the organism in the absence of external perturbing factors, the rhythm is said to be endogenous. As an example we may use the sleep movements of the first leaves of the common bean. In the daytime, the blades are more or less horizontally placed, and the petioles are spread wide from the stem. At night, however, the leaf blade assumes a vertical orientation, and the petioles assume a more acute angle to the stem. The movements can be visualized by connecting the leaf tip to the writing stylus of a rotating kymograph drum by a fine thread. If such a plant is placed in a continuously darkened, thermostated room, the oscillation in leaf orientation continues for several days, until starvation sets in (Fig. 74).

What is the physical basis for such a rhythm? Although we do not yet have a very clear concept of the answer, we do know that the duration of the rhythmic oscillation is temperature-independent, yet sensitive to various agents that interfere with normal cell metabolism. These facts greatly limit and greatly complicate the kinds of schemes that may be proposed to explain rhythms.

Certain researchers have proposed that endogenous rhythms are

Fig. 74. A recording of the endogenous movements of bean leaves.

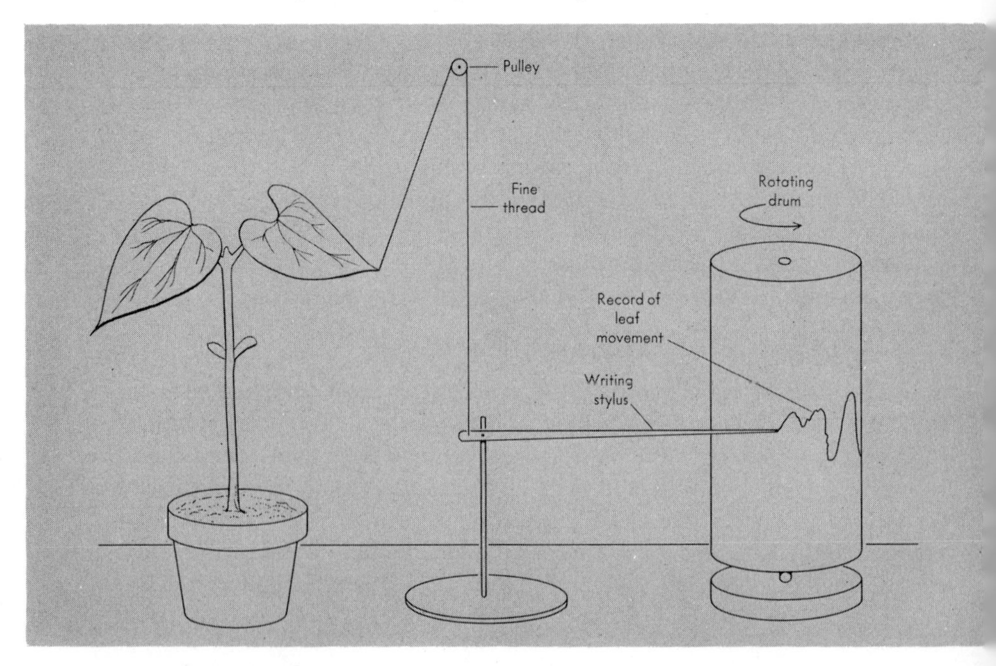

intimately related to the photoperiodic system. For example, those varieties of soybeans that have marked endogenous leaf movements have specific short-day requirements, while those without marked rhythms tend to be day-neutral. Also, the efficiency of a light flash given during a prolonged dark period in enhancing or inhibiting flowering varies with the time of day it is given, and the pattern of variation seems to describe a 24-hour cycle.

Whether or not rhythms are involved in photoperiodism, they are sufficiently interesting to merit extended study. Rhythms are found in all types of organisms, including microorganisms, higher plants, and higher animals, and they may well provide vital new information on the nature of biological systems. Since certain microorganisms and animals are also photoperiodically sensitive, comparative studies in this area are also warranted. Indeed, in the study of endogenous rhythms, as with all biological problems, experimentation with any biological system may be expected to yield information of potential relevance to all cells.

Selected Readings

GENERAL

Bonner, J., and A. W. Galston, *Principles of Plant Physiology*. San Francisco: Freeman, 1952. A brief introduction to the physiology of the green plant, with emphasis on biochemistry, growth, and development.

Meyer, B. S., D. B. Anderson, and R. H. Böhning, *Introduction to Plant Physiology*. Princeton: Van Nostrand, 1960. A general textbook of plant physiology. Especially complete treatment of water relations and plant nutrition.

THE CELL

Butler, J. A. V., *Inside the Living Cell*. New York: Basic Books, 1959. A lively, semi-popularized account of the nature of living systems.

Giese, A. C., *Cell Physiology*. Philadelphia: Saunders, 1957. A general textbook that prepares the serious student for future studies.

Swanson, C. P., *Cytology and Cytogenetics*. Englewood Cliffs, New Jersey: Prentice-Hall, 1957. A scholarly account of cellular structure with especially detailed discussions of chromosome make-up and behavior.

PLANT ANATOMY

Esau, K., *Plant Anatomy*. New York: Wiley, 1953. A beautifully written and illustrated work. Clearly the best in its field.

PLANT NUTRITION

Kramer, P. J., *Plant and Soil Water Relationships*. New York: McGraw-Hill, 1949. A detailed and technical account of the physiology of water in the higher plant.

Hoagland, D. R., *Lectures on the Inorganic Nutrition of Plants*. Waltham, Mass: Chronica Botanica, 1948. A series of lucid lectures on mineral nutrition by the leader in the field several decades ago.

Whittingham, C. P., and R. Hill, *Photosynthesis*. New York: Wiley, 1955. A brief treatment of the biochemistry of higher-plant photosynthesis.

Arnon, D. I., "The Role of Light in Photosynthesis," *Scientific American*, 203 (1960), 104–118. A brief, popular review of the discovery of photophosphorylation and its significance.

PLANT HORMONES

Went, F. W., and K. V. Thimann, *Phytohormones*. New York: Macmillan, 1937. A highly readable synthesis of the early literature on plant hormones by the discoverer of auxins and a distinguished colleague.

Leopold, A. C., *Auxins and Plant Growth*. Berkeley: University of California Press, 1955. A useful supplement to the previous reference. Brings the field up to date as of the middle 1950's.

Audus, L. J., *Plant Growth Substances*. London: Leonard Hill, 1959. A well-written, scholarly, detailed treatment of the massive literature on this subject.

GROWTH AND MORPHOGENESIS

Sinnott, E. W., *Plant Morphogenesis*. New York: McGraw-Hill, 1960. A competent, modern treatment of some fascinating problems in the development of the higher plant.

Borthwick, H. A., and S. B. Hendricks, "Photoperiodism in Plants," *Science*, 132 (1960), 1223–1228. A brief, semipopular account of the development of the phytochrome story.

Index

Index

A

B

C

D

E

F

G

H

I

K

L

M